The Self-Organizing Soot

SHOAIB AHMAD

Copyright © 2022 Shoaib Ahmad

All rights reserved.

ISBN: 9798812728915

DEDICATED

To

Nilore

who watched with wonder
the slow gazing River Soan
swallowed by Lake Simly

Then

it re-emerged from Lake Simly

And Now

Nilore
delivers lectures on

Self-Organization

CONTENTS

	Acknowledgments	vi
1	Carbon Clusters, Cages and the Soot	8
2	The Regenerative Soot	29
3	A Continuum Elastic Model of Nano Curvature	59
4	The C_2 Gas	85
5	The Degenerate Fermi Gas of Fullerenes	110
6	End-Directed Emergence	128
7	The Dynamic Emergence	156
8	The Information Manipulating Soot	177
9	List of References	191

ACKNOWLEDGMENTS

The author is indebted to S.A. Janjua, S. Javeed, S. Zeeshan, A. Ashaf, S.D. Khan and K. Yaqub, for participation in the numerous experiments planned, executed and reported in the latter half of the book.

M. Yousuf, M.S. Abbas, B. Ahmad, A. Qayyum, M.N. Akhtar, W. Arshad, T. Riffat, M. Ahmad, R. Khalid, A. Aleem, A. Ellahi, M. Uzair and S. Hussain participated in the work reported in the first 2 chapters. Their participation is gratefully acknowledged.

The experimental research reported in this book was conducted at the indigenously developed CPA laboratory at PINSTECH (Islamabad) and at 2 MV Pelletron laboratory at GC University (Lahore). The author expresses his gratitude to the technical staff of the two laboratories.

The SRS model described and employed here, was first presented, by the author, in an extended discourse, a Course given at PIEAS, 'Fractals and Irradiated Solids' in the spring of 2018. The PINSTECH and PIEAS colleagues and research students are acknowledged for their enthusiastic support and helpful discussions during the Course.

The book would not have been written or completed without the encouragement and support of Khalida Ahmad; the author expresses his profound gratitude.

The Self-Organizing Soot

Chapter 1

Carbon Clusters, Cages and the Soot

1.1. The self-organizing Soot

In his Nobel lecture, Smalley described the discovery of the Buckyball-C_{60} as the emergence of 'order out of disorder' [1]. Smalley was referring to the highly symmetric structure of the family of the closed cages of carbon, classified as fullerenes, which emerged out of the apparent disorder of the laser-ablated, high temperature soot and self-organized. The paper that earned Kroto, Curl and Smalley their Nobel prize in Chemistry, was entitled 'C_{60}: Buckminsterfullerene' [2]. Theoretical chemists showed that the closed cages of carbon- the fullerenes had to fulfil two basic conditions; the first implied that each fullerene must have exactly twelve (12) pentagons while the number of hexagons could vary and the second implied that the addition of hexagons would induce isomeric variability. Therefore, the smallest fullerene was predicted to contain twenty (20) C atoms as a dodecahedron of twelve (12) pentagons with no hexagon. This smallest fullerene would have only one isomer. All fullerenes $\geq C_{24}$ have multiple isomers. C_{60}, for example, has 1812 isomers in which the hexagonal positions change around the twelve, omnipresent pentagons [3]. These two structure defining conditions suggested that the laser ablated plume of soot should contain a large range of fullerenes with variable number of C atoms. It also implied that multiple isomers of C_{60} must be formed in the grand canonical ensemble of the wide range of fullerenes ($\leqq C_{60}$) with their respective isomeric variability. The emergence of the perfectly symmetric icosahedral cage with the twelve, non-abutting pentagons was indeed 'order out of disorder'. The non-abutting pentagons ensured the stability of the sp^2-bonded spherical configuration with the uniformly distributed steric strain that stabilized the cage [3]. Smalley had implied the operation of the three stages of the emergence of 'order out of disorder'; the first was the formation of the closed cages in the ablated, disordered soot, the second implied self-organization of the larger cages through the cage-to-cage transformations and, the third and the most conspicuous one demonstrated that only one out of the 1812 isomers of C_{60} cages, with the non-abutting pentagons, would emerge out of the ensembles of a large number of fullerenes > C_{60}. The title 'C_{60}: Buckminsterfullerene' in their epic Nature paper [2] implied these three, essential, vital steps and stages of the self-organizing soot.

Smalley group's experiments on the carbon soot produced by laser ablation of graphite contained carbon clusters which included linear chains, rings and closed cages of variable sizes and isomers. Their experiment indicated that the manipulation of experimental conditions of the soot had significant effects on the net outcome. They successfully managed to create the conditions where only the cages with the highest symmetry survived. The conditions and the criteria for the self-organizing soot transforming into closed cages with the highest symmetry will be discussed in Chapter 3.

Soon after the discovery of fullerenes, another experimental observation of self-organization amongst the ensembles of the sputtered and dislocated carbon atoms was reported in the high energy electron microscopy by Iijima [4]. He reported that the knocked off and displaced atoms from graphite re-organized in the form of nanometer sized cylinders called carbon nanotubes. He observed that the intense beams of high energy electrons could dislodge C atoms from their sites that regrouped as cylindrical tubules of nanometer size. Both single- and multi-walled carbon nanotubes were produced and reported. Carbon nanotubes and fullerenes added to Carbon's allotropic forms of graphite and diamond. Similarly, the newer and more diverse experimental techniques started being reported from all over the world and a new era of research on Carbon nanostructures began. In Germany, Kratchmer's group was conducting research on C clusters. Once fullerenes had been reported as Carbon's new allotropic form, they demonstrated that the dc arc discharge between graphite electrodes in the presence of noble gases was capable of producing fullerenes in larger quantities as compared with laser ablation [5]. All one needed was, create carbon vapor or some form of soot and various shapes and forms of carbon cages would be formed.

Since the discovery of fullerenes and C nanotubes, various methods and techniques have been employed in addition to the laser ablation, intense high energy electron beam irradiation and the dc arc discharge. The overriding objective has been to generate and investigate the mechanisms of the Carbon soot and the broad spectrum of the C nanostructures including the linear chains, rings, sheets and the closed cages. The ability to produce and modify the nanometer sized-cages' structures and properties initiated active research, in which scientists belonging to multiple disciplines that included chemistry, physics and material sciences, conducted and reported a broad range of activities [6].

During the Nilore experiments in the mid-nineties, the Carbon cages-specific serendipity had its way during the Direct Recoil experiments on the cluster emission from the heavy ion irradiated graphite [7-9]. In addition to the Carbon atoms recoiling off the surface, the whole series of clusters were also detected in the Direct Recoil spectra. These experiments suggested that ensembles of C clusters and closed cages formed on the irradiated graphite surface. Their formation and fragmentation sequences occurred as a function of the heavy ion's energy, angle of incidence and the ion doses. The mechanisms and processes of the heavy ion-induced formation and manipulation of the soot in the form of C clusters, the open and the closed cages on the irradiated graphite surfaces are described in the succeeding sections.

1.2. Carbon cluster sputtering

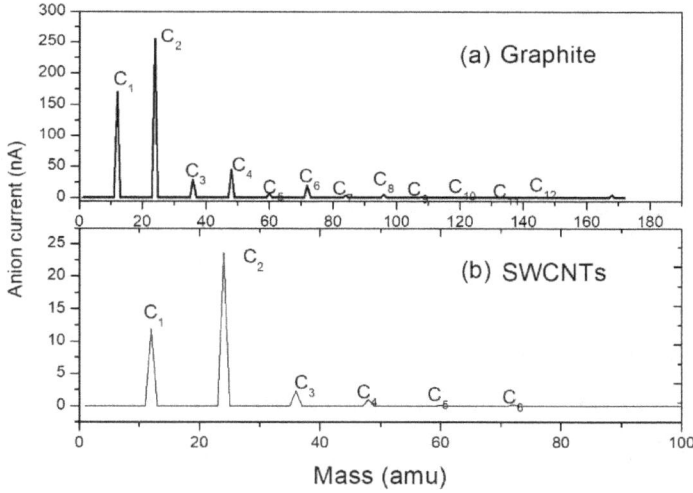

Fig. 1.1. Mass spectra of negative atoms and clusters C_x; $x \geq 1$ sputtered as anions from (a) graphite and (b) SWCNTs irradiated with Cs^+ ions with 5 keV. Data from ref. [10].

The experimentally observed phenomenon of the formation of Carbon soot on irradiated graphite surfaces, in vacuum, can be explained by the mechanisms of emission or sputtering of the multi-atomic C_2, C_3, C_4 and the higher C clusters. Monatomic C_1 is present but with relatively smaller number densities as compared with the C_2 [10]. The typical mass spectrum from Cs^+ irradiated graphite is shown in Fig. 1.1 emitted from the Source of Negative Ions by Cesium Sputtering (SNICS) [11]. It contains an array of the negatively charged anions including the monatomic C_1. Figure 1.1 shows a set of the two mass spectra of sputtered C anions with Cs^+ irradiations at energy $E(Cs^+) = 5.0$ keV. The sputtered anions' spectra are shown for the irradiated graphite and the single-walled carbon nanotubes (SWCNTs). Prior to the discovery of fullerenes, mass spectrometric studies on the thermally subliming and evaporating graphite targets had indicated that the typical emitted species consisted of the C clusters up to thirty (30) C atoms [12-14]. These studies had also demonstrated the advantages of detecting negatively charged clusters anions (C_x^-) over the corresponding positively charged cations (C_x^+). Majority of the sputtered species are emitted as neutrals from the irradiated surfaces. That makes it more efficient to attach an electron to a neutral cluster than to remove one. Ionization may lead to the cluster fragmentation in addition to the charge removal. In the case of anions, the electron affinities are ~1-3 eV for most clusters, while ionization energies (≥ 10 eV) are ~fragmentation energies [15,16]. In addition, the sputtering of graphite in SNICS has the advantage of the choice of the range of irradiation energies $E(Cs^+)$ between 0.1-5.0 keV. Fig. 1.1(a) has the mass spectrum from irradiated graphite powder and Fig. 1.1(b) is from SWCNT under similar irradiation conditions. An order of magnitude higher anion current densities from the 3-D solid target (graphite) in Fig. 1.1(a) can be noticed as compared with those from 2-D SWCNTs in Fig. 1.1(b).

Fig. 1.2. has the normalized number densities or probabilities of emission p_x of the first four anion species C_1, C_2, C_3, C_4 sputtered from (a) graphite, (b) SWCNTs and (c) MWCNTs as a function of $E(Cs^+)$ between 0.5-5.0 keV. Diatomic carbon C_2 is the most intense sputtered species from the three allotropes of Carbon. Only in the case of graphite, monatomic C_1

has the second highest number density, while clusters (C_3, C_4) dominate the mass spectra from the nanotubes along with C_2. Clusters dominate the sputtered species' spectra from all of the irradiated carbon allotropes. The mass spectrometric observation of the preponderance of multi-atomic clusters over the sputtered monatomic C_1 indicates that the nonlinear thermal spikes are effectively operating in addition to the linear, binary atomic collision cascades. Radiation-induced damage in Carbon's allotropes is different from its counterparts in metallic solids where the binary atomic collision cascades determine and define the nature of the damage while the nonlinear, thermal processes dominate the irradiated graphite and nanotubes.

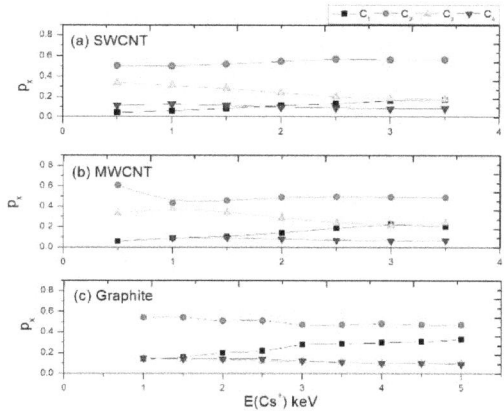

Fig. 1.2. The normalized probabilities of the sputtered species p_x of the monatomic C_1 and clusters C_2, C_3 and C_4 plotted as a function of the energy of the irradiating ion $E(Cs^+)$ for (a) SWCNTs, (b) MWCNTs and (c) graphite. Data from ref. [10].

1.3 Thermal origin of cluster sputtering

The linear, binary atomic collision cascade (CC) sputtering theories [17] do not explain the sputtering of clusters. These are theories of the spreading of cascades of binary atomic collisions that result in the generation of the recoiling atoms, creation of single vacancies and sputtering of the atomic constituents from the irradiated surfaces. A collision cascade theory predicts sputtering yields. It utilizes the ion to target mass ratio, the angle of ion incidence, interatomic potentials and the energy of the incident ion as the parameters. The irradiating ion's energy is shared among the target recoils in binary collisions creating primary recoils, the secondary and the tertiary recoils of C atoms that form the CC. Sputtering yield, therefore, counts the number of recoiling target atoms that leave the outer surface per ion. The collision cascade-based theories of sputtering [17] can explain majority of the experimental data from metallic targets where the monatomic sputtering yield is proportional to the energy deposited in the linear cascades of the recoiling target atoms. The nonlinear thermal spikes were introduced to explain the thermal origin of a fraction of the sputtered particles with small energies~ a fraction of eV [18-25]. Experimental evidence for thermal spikes grew over the years [23-25]. Majority of the experiments, theoretical and computational modelling of nonlinear sputtering dealt with the 3-dimensional materials. The irradiated solid's dimensions were generally larger than the range of the bombarding ion. On the other hand, the mono-layered, 2-dimensional sheets of sp^2-bonded C atoms share and dissipate energy differently as compared with their 3-D metallic counterparts. The irradiated graphene sheets and the nanotubes of carbon restrict the number of direct recoils into the single planes, thus making sputtering by collision cascades less efficient. At low recoil energies, near the end of the cascades, higher percentages of energy are deposited in phonons that raise the local temperature. Localized thermal spikes are generated. The overall effect is the lower contribution of the monatomic species and higher for the clusters, as shown in Figures 1.1. and 1.2.

The emission of clusters can be described as the net outcome of the emergence of a locally subliming region as a localized thermal spike (LTS) [26]. The LTS-model utilizes the probabilities of emission of clusters associated with the formation energies of the respective multi-vacancies and the sublimation temperature T_{Sub}. The model was developed for the monolayer containing N carbon atoms of a planar graphene sheet or the SWCNT, but it is applicable to the bulk solids as well [10]. The monatomic or multi-atomic vacancies are created if one, two, three, four or x- numbers of carbon atoms are removed in the form of clusters C_x. Single carbon atoms or clusters C_x are bonded to the matrix of the surrounding C atoms with their respective binding energies E_x. The n vacancies with x C atoms in a target with N carbon atoms, are created in $W = \frac{N!}{(N-n)!n!}$ ways. The associated entropy is $S_x = klnW$. The internal energy $U = nE_x$ and temperature T_{Sub} are related to E_x and leading to entropy S_x through $1/T_{Sub} = \frac{k}{E_x}(\partial W/\partial n)$. This relation leads to the probability of emission of an x-numbered cluster [26]

$$p_x = n_x/N_S = \{(\exp(E_x/kT_{Sub}) + 1\}^{-1} \qquad \text{eq. (1.1)}.$$

Here p_x denotes the probability of creation of a vacancy of the emitted x-C atoms. The normalized yields of the sputtered clusters can be obtained from the experimental mass spectra of clusters C_x. The experimental density for the emission of C_x is directly proportional to the probability of the thermally created vacancies ($\propto p_x$). The probabilities of emission of C_2, C_3, C_4 and the higher ones, is $p_x = n_x/N_S$, where N_S is the total number of C atoms in the spike region. Since N_S is an unknown quantity, therefore, the ratio of any two probabilities can be used to eliminates N_S. The ratio of the probabilities allows calculation of the LTS temperature by using $p_x/p_y = \frac{n_x}{n_y} = \{exp(E_y/kT_{Sub}) + 1\}/\{exp(E_x/kT_{Sub}) + 1\}$.

The spike temperature T_{Sub} is obtained as [10,26]

$$T_{Sub} \cong [(E_{xv} - E_{yv})/k)][\ln(p_y/p_x)]^{-1} \qquad \text{eq. (1.2)}.$$

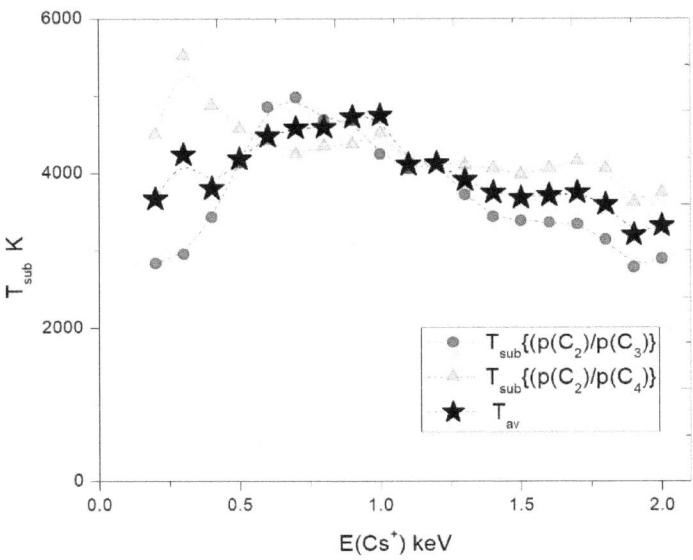

Fig. 1.3. Localized thermal spike temperature *Tsub* calculated from the normalized densities of C_2/C_3 and C_2/C_4 against $E(Cs^+)$. Average for the two ratios of clusters, *Tsub*(C_2/C_3) and *Tsub*(C_2/C_4) shown as bold black starred T_{av}~4000 ±10% K. [26].

The values of the energies of formation of single and double vacancies E_{1v} and E_{2v} used in the LTS model were obtained from the DFT and DFT-based TB calculations for the SWCNTs [27-31]. The average values of tri- and quarto-vacancies obtained at the temperatures obtained from C_2/C_1 are E_{3v}=5.13±0.5 eV and E_{4v}=5.73±0.5 eV for the $E(Cs^+)$ = 0.4 to 1.0 keV. From the formation energies of C_2, C_3, C_4 and their experimental probabilities, two sets of temperatures are calculated and shown in Fig. 1.3. Along with the average T_{av} [26].

The fundamental difference between the energy dissipation mechanisms in bulk solids and graphene is due to the covalently bonded C atoms on the vertices of the hexagons on a mono-shelled surface. Fig. 1.4. shows a hexagonal C-network struck by a Cs^+ ion. The first struck atom numbered **#1**, receives energy $E_1 \equiv 1$. For $E_1 \gg E_{dis}$ where E_{dis} is the energy to remove a C atom from its lattice site, a typical binary collision cascade is generated with the energetic recoiling atom. The collisions proceed until the recoiling energy $< E_{dis}$. Such collisions generate Frankel pairs of interstitial atoms and vacancies. However, the concept of interstitials is strictly valid only in bulk solids, these become the sputtered atoms in the case of graphene sheets and the SWCNTs.

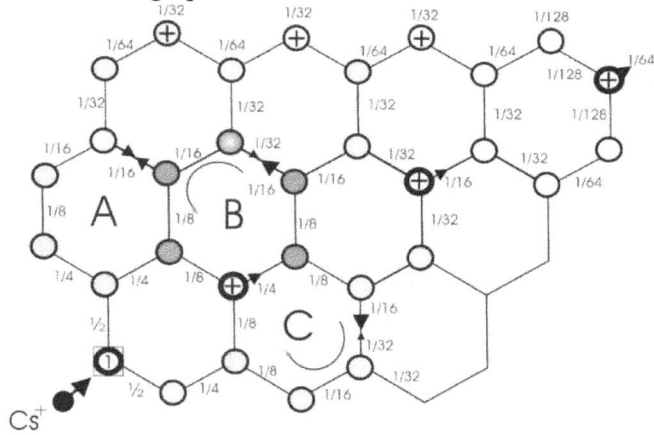

Fig. 1.4. The initiation of the localized thermal spike in a 2-D graphene sheet. C atoms on the hexagonal vertices' network shown with the first struck atom labelled as **1** (circled & boxed). Subsequent energy sharing in binary collisions is shown with the received energies along the bonds (as a ratio of the recoil energy). Energy recycling is shown in the rectangles A, B and C.

For energies $\lesssim E_{dis}$, three different processes of energy sharing exist among the atoms of the hexagonal network. The first is the spreading of the recoil **#1** outwards and away from the point of the initial impact. The second mechanism is shown by the opposing arrows, like a bow-tie, in the hexagon labelled A. The atoms on the opposing edges of the hexagon receive equal energies, in opposite directions. The collisions increase the vibrational frequencies by localizing the energy in collisions with the surrounding atoms. This may lead to the localizing of deposited energy. The third mechanism is shown by two unequal, opposing arrows on the interatomic bonds in hexagons B and C. These are due to collisions between atoms which deliver, or share, different amounts of energies, like 1/16 and 1/32 in the two cases. The excess energy is directed towards the atom with lesser energy. The energy recycling in hexagons B and C is shown by the curved arrows. The latter two processes increase the vibrational energies and consequently the local temperature. It eventually generates local disorder with average atomic energies~1/5-1/3 eV. The spike temperatures T_{Sub}~4000±10% K can be generated from the sputtered cluster probabilities [10, 26]. At these temperature clusters are emitted from the localized subliming regions. Fig.1.5 is the graphical presentation of the process of recycling of the residual energy within a set of seven adjacent hexagons, through the energies shared by C atoms on the vertices.

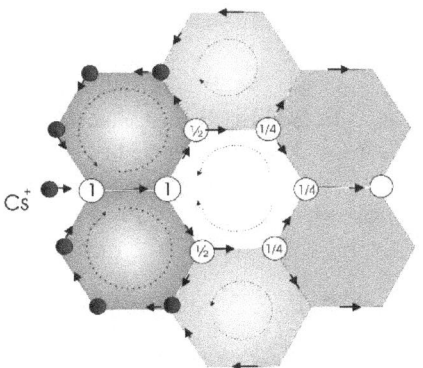

Fig. 1.5. Recycling of energy in hexagons, induces higher vibrational temperature. Denser colors of the hexagons indicate the rising temperatures leading to the LTS.

1.4. Direct Recoil Spectroscopy of Clusters and Cages

The evolving clusters and their fragmentation under continuous ion bombardment of graphite are detected in the energy spectra of the Direct Recoils (DRs). DRs are emitted as a result of collisions between the incident ions and the re-constituted surfaces. The successive DR spectra reveal the sequences of the energetics of the C–C bond formation of the clusters and cages that have been formed and deposited on the surface. In this section, DR spectroscopy is described for understanding the sequential progress of the emerging soot as the in situ experimental evidence [7-9].

The constituents of a surface recoiling in a binary collision with an incident ion are the primary knock-ons of the radiation damage theory also known as the Direct Recoils-DRs. A DR carries a characteristic energy which is a function of the target to projectile mass ratio (m_2/m_1),

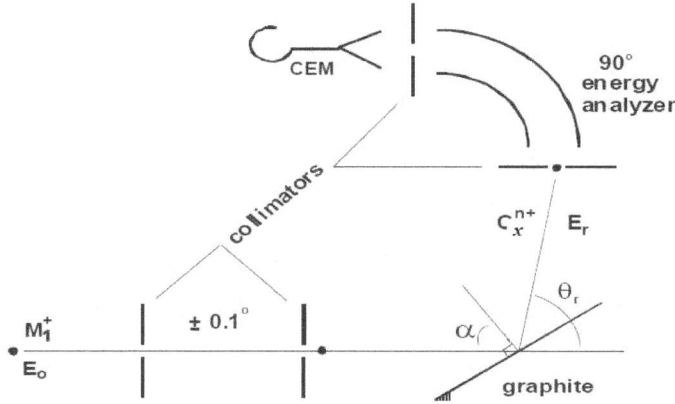

Fig.1.6. The Nilore DRS experiments utilized an indigenously designed and fabricated 250 keV heavy ion facility. Ar^+; Kr^+ and Xe^+ beams of > 1 mm diameter with energy between 50 to 250 keV delivered to a target 2 meters from the end of the Accelerator tube. The facility equipped with a hollow cathode duoplasmatron operated at pressure ~ 10^{-2}–10^{-3} mbar, with the accelerator delivering a few μA collimated beam with ~ ± 0.1° divergence on the target. The pressure in the target chamber ~ 10^{-7}–10^{-8} mbar. [ref. 7].

The angle of recoil θ_{DR} and the ion bombarding energy E_0. The energy of graphite atoms of mass m_2 recoiling at angle θ_{DR} is given by Bohr [32] as

$$E_{DR} = 4\frac{m_1 m_2}{(m_1+m_2)^2} E_0 \cos^2 \theta_{DR} \qquad \text{eq. (1.3)},$$

where m_1 and E_0 are the mass and energy of the incident ion-the projectile. The differential cross section $d\sigma_r/d\Omega$ of the recoiling particles, in the Lab frame, can be worked out from the differential scattering cross section $d\sigma(\zeta)/d\omega$ in the centre of mass (C.M.) system for C.M. angle as $d\sigma_r/d\Omega = 4\sin(\zeta/2)\, d\sigma(\zeta)/d\omega$ [32]. By using Kr–C potential and following ref. [33], $d\sigma(\zeta)/d\omega = \frac{m_2 E_0}{(m_1+m_2)} \frac{3.05 Z_1 Z_2}{\left(Z_1^{1/2}+Z_2^{1/2}\right)^{2/3}} \frac{\pi^2(\pi-\zeta)}{\zeta^2(2\pi-\zeta)^2 \sin\zeta}$.

For a cluster recoiling from the sooted surface of graphite after ion-cluster collision, the $d\sigma_r/d\Omega$ can be calculated for selective values of large recoil angles $\theta_{DR} = (\pi/2 - \zeta/2)$.

For the heavy ion (Ar^+, K^+, Xe^+) irradiation, the conditions in which the minimum direct recoil energy $E_{DR}^{min} \sim 10\ eV$ corresponding to large recoil angle $\theta_{DR} \approx 89°$ are to be established. This minimum is necessary for the detection of the largest detectable clusters as Direct Recoils. In DR spectroscopy, the heavy clusters receive successively smaller energies. The experimental setup is shown in Fig. 1.6. where the beam and the recoil particles' collimators with less than $\pm 0.1°$ divergence. Channel Electron Multiplier (CEM) was used for the charged cluster detection. CEM with typical gain of $\sim 10^7$ fed the recoiling species' data to a PC. The energy analyzer's condenser plate potential was increased in variable steps through a function generator. The electrostatic analyzer ensures the detection of up to tens of keV heavy recoiling clusters which implies all clusters, from the smallest to the largest, can be detected. This aspect of DRS is advantageous since the corresponding mass spectrometry via momentum analysis would be limited by the requirement of huge dipole magnets. Although momentum analysis of clusters is desirable to unambiguously characterize the m/q values but the required magnetic fields become unrealistically large for experimental arrangements of DRS. For example, in case of $\theta_{DR}= 79.5°$, a large magnet is needed with $B_0\rho =$ 4 [T-m] for resolving clusters e.g., C_{60}. Higher C cages would require even larger magnetic fields. The DRS, during the ongoing continuous ion irradiation, is capable of providing the sequential evidence of the emerging, fragmenting and the surviving clusters and cages that constitute the soot. The evolution of soot on the irradiated graphite surfaces was monitored in these DRS experiments.

1.5. Direct-recoiling clusters at $\theta_{DR} = 87.8°$

Fig.1.7. shows the energy spectra of direct recoils from 100 keV Kr^+ ion bombarded graphite at recoil angle $\theta_{DR} = 87.8°$. Three consecutive spectra are shown with 4 μA beam incident at grazing incidence with respect to the surface normal $\alpha \sim 80°$. The first spectrum (Fig. 1.7(a)) is taken after a dose of 2.5×10^{13} ions. The entire fullerene range is present with a broad peak around C_{70}^+ to C_{50}^+. Higher x as well as the lighter ones including $x < 36$ and the linear/chain structural combinations ($x = 1$ to ~ 10). The next spectrum Fig.1.7(b) has C_{80}^+ as the dominant species while other cluster especially the fragments C_{x-2}^+ resulting from C_2-spitting sequences $C_x \rightarrow C_{x-2} + C_2$ leading to C_{80}^+, C_{56}^+, C_{44}^+, C_{40}^+ that are conspicuous by their relative abundance. The lower order fullerenes show increased share of the total yield. C_{70}^+ and C_{50}^+ are present but C_{60}^+ is not significantly present. The gradual building up of the C_{50}^+ and its fragments ($C_{48}^+, C_{46}^+, C_{42}^+, C_{40}^+$) can be seen from Fig. 1.7(c). The smaller clusters are present but their total yield is much smaller than that of the higher (i.e., $> C_{50}^+$) fullerenes.

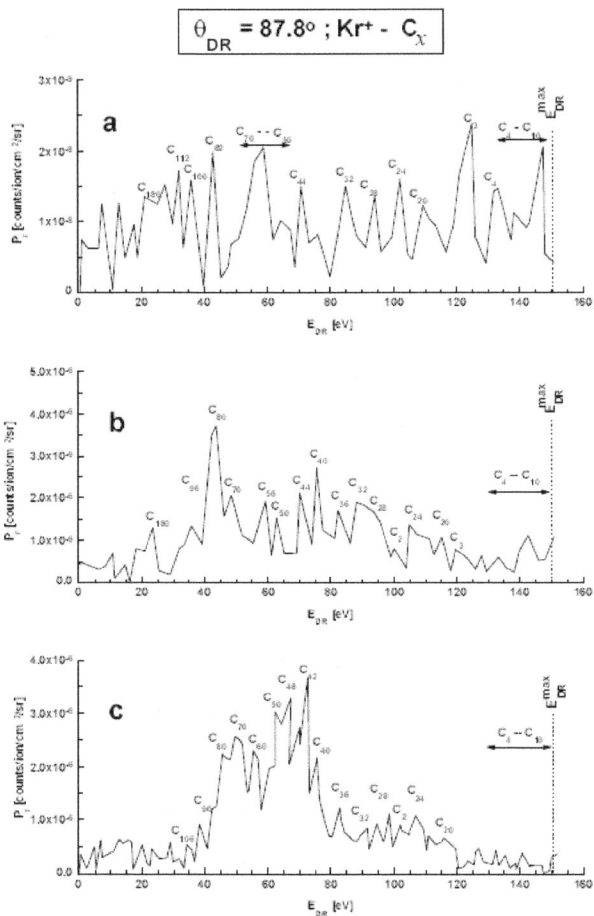

Fig.1.7. Shows the Direct Recoil energy spectra from Kr$^+$-C at 100 kV of the Clusters C$_x$ are seen recoiling with characteristic energies E_r. Fig. 1.7(a-c) present the sequences of dynamic formation and gradual fragmentation of various clusters. E_{DR}^{max} is the maximum transferable energy to a cluster in a direct recoil.

The range of the DR energies at $\theta_{DR} = 87.8°$ for 100 keV Kr$^+$−C for various C_m^+ is between 48 and 147 eV. This is about 22 times less than their corresponding values at $\theta_{DR} = 79.5°$. For Xe$^+$−C, these recoil energies are even lower. The large recoil angle (~90°) arrangement has two important aspects for the detection of charged clusters; (a) due to the much smaller recoil energies, the differential cross section is enhanced for the heavier projectiles, by at least an order of magnitude, and (b) the monomer multiplicity is considerably reduced as the slower, larger cages pick up electrons while leaving the surface.

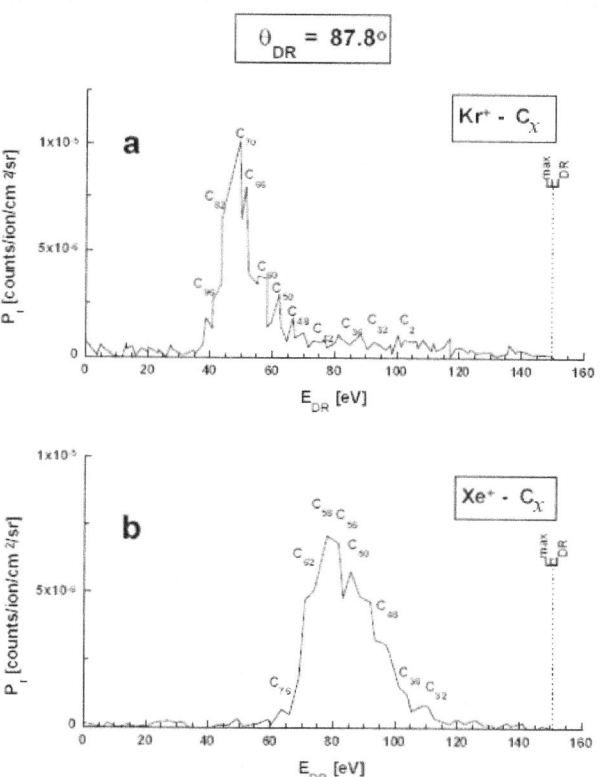

Fig.1.8. DR spectra of the self-organizing clusters from (a) 100 keV Kr^+ and (b) Xe^+ irradiated graphite. These are the end-of-sequence spectra of the emerging and surviving cages that resulted from the cumulative, increasing ion flux. Fig. 5 of ref. [9].

Fig. 1.8(a) and (b) are for the DRs sputtered off the heavily irradiated target at $\theta_{DR} = 87.8°$, with 100 keV Kr^+ and Xe^+, respectively.

Both of these spectra were obtained after $\sim 10^4$ s of ion bombardment at grazing incidence with the surface $\alpha = 80°$ which implies sputtering away of 10^3–10^4 surface layers. Fig. 1.8(a) for Kr$^+$–C is peaked around C_{70}^+ and its fragments C_{68}^+ and C_{66}^+ along with the larger clusters on the shoulders of C_{82}^+-to-C_{96}^+ and the tail of smaller clusters with reduced intensities. The Xe$^+$–C spectrum of Fig.1.8(b) is peaked around by C_{60}^+ with smaller cages on its tail. The peak is surrounded by C_{50}^+ and C_{62}^+ are present on the shoulders [9].

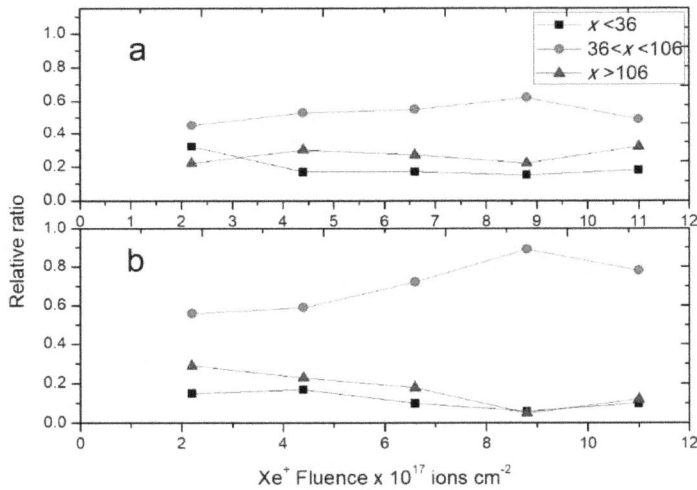

Fig. 1.9. The heavier clusters C_m ($m > 106$) show a gradual increase in the DR spectra of the cages recoiling from Kr$^+$ and Xe$^+$ irradiated, sooted surfaces.. The group of clusters with $36 \leq m \leq 106$ is slowly increasing as a function of the dose and the individual cluster contributions do not show significant variations. At higher doses, the cage formation and fragmentation processes result in the higher yields of the range of cages with $36 \leq m \leq 106$. From ref. [8,9]

Fig. 1.9. summarizes the relative contributions of the various carbon clusters C_x^+ in 3 main groups; (1) the largest closed structures with $x > 106$, (2) the main group centered in the regime $36 < x \leq 106$ and (3) the smaller fullerenes with x < 36. The evolution of the larger closed cages becomes statistically significant with Kr⁺ irradiation.

It must be emphasized that the soot formed on the irradiated surface is recycled by the ongoing, continuous interactions with the high energy Kr⁺ and Xe⁺ ions. The sooted surface is not enclosed in a high temperature oven or plasma. The thermal spikes are localized around the ion tracks in the soot on the irradiated surfaces. The comparison with the enclosed surroundings with be discussed in Chapter 2 in the context of the regenerative soot.

1.6. The Summary of the DR spectroscopy and the mechanisms of the emerging soot

The heavy ion induced physical and chemical processes in the irradiated graphite lead to the formation of clusters and closed cages and to their subsequent fragmentation. The continuous, prolonged irradiation has three essential features that made it useful for the study of the formation of C clusters and the closed caged fullerenes:

(1) Almost all existing bonds between C atoms along the ion path are broken on the time scale $\sim 10^{-14} - 10^{-15}$ s. For example, a 100 keV Xe⁺ ion has a range of ~ 660 Å and takes $\sim 10^{-13}$ s to deposit its energy before coming to a stop. The primary recoil energy distribution $\propto E_r^{-3/2}$. For Xe⁺–C_1 collisions, the C recoil energy $E_r(C_1)$ varies from $\sim E_b$ (binding energy) at $\theta_r(max) \approx \pi/2$ to 15.4 keV for $\theta_r(min) \approx \pi/4$. These values correspond to the range of the projectile's scattering angles between $\varphi_{min} \sim 0.1°$ and $\varphi_{max} \{sin^{-1}(m_2/m_1)\} = 5.24°$. Since the majority of these ion-target atom collisions favor low energy recoils, the forward moving recoils in these cones have half angles $\theta_r > \pi/4$ in the case of Xe⁺–C_1. The cone density along the track follows from the primary

recoils' energy distribution $\propto cos^3_{\theta_r}$. Thus the lower E_r and high θ_r primary atomic recoils are wrapped around the ion path with almost constant linear density.

(2) The energy of these primaries is further distributed in the secondary and tertiary C_1–C_1 collisions with the characteristic scattering and the recoil angle = π/2 leading to collision cascades with the linear recoil density $\propto E_r^{-2}$ [18,34]. In the heavy ion bombarded metallic targets, the characteristic time scales of the cascade $\tau_{cascade} \sim 10^{-12}$ s [19-22]; for the irradiated insulators and semiconductors it typically takes an order of magnitude longer time.

(3) Direct Recoils exclusively deal with the emission of the outer surface constituents. Their sequential profiles contain information that can be used to characterize the dynamics of the soot consisting of the clusters and cages' and the subsequent fragmentation under the continuous ionic bombardment.

For the series of DRS experiments reported above, the irradiating ion energy, its angle of incidence α with the surface normal and the direct recoil angles of the emitted species θ_r, were chosen such that the heavy ions deposit considerable amounts of energy ~40 − 160 $[eV/Å]$ in a restricted volume of not more than 30 atomic layers beneath the irradiated graphite surface. It required choosing $\alpha \geq 80°$ with the ion penetration depth ~ 660–1300 Å at 100 keV. The incident heavy ion flux ensured that in addition to the generation of intense collision cascades, maximum randomization of the target constituents may occur.

The changes in the surface topography of the heavily irradiated graphite are directly related to the rather high sputtering yield (20-40 atoms/ion) for the energetic heavy ions. The grazing angles of ion incidence ~ 10° between the ion beam and the surface, further reduces the effective penetration depth transverse to the surface thereby restricting the damage to ~ 100 Å beneath the surface. The sputtered species are emitted as a result of the collision cascades intersecting the surfaces, and the direct recoiled species are related with the dynamically emerging surface constituents. Such surfaces go through the surface re-building processes due to the re-deposition of sputtered species emitted with very small energies. Typical values of the sputtering yield per unit solid angle $dS/d\Omega$ for 100 keV $Xe^+ \rightarrow C$ are $\sim 10^6\, C_1/Xe^+/st.rad$. Comparison of this value can be done with $dP/d\Omega \sim 10^{-9} C_m/Xe^+/st.rad$ from the data presented in e.g., Fig. 1.7 and 1.8. The fifteen (15) orders of magnitude difference in the rates of the two ion induced emission mechanisms clearly shows that while sputtering is the main agent for the topographical changes, the DRs carry the instantaneous surface constitution information. The DRs that originate in the binary collisions between incident ions and the surface constituents carry with them the information that characterizes the dynamics of cluster formation and fragmentation within the reconstructed surface.

The Self-Organizing Soot

Chapter 2

The Regenerative Soot

2.1. Introduction

Chapter One introduced the experimental observations of the soot resulting from the C clusters sputtered from graphite and nanotube surfaces under heavy ion irradiation. The mechanisms and processes of the regeneration of soot were also presented under the continuous irradiation detected by DR spectroscopy. In this chapter, the regeneration of the soot is presented in the context of a specially designed source with graphite hollow cathode and hollow anode discharges. The source will demonstrate that the environment of regeneration of the soot containing a wide range of clusters and cages is ideal for investigating the continuous, ongoing formation and fragmentation of C clusters. For the mass spectrometry of the clusters an $E \times B$ velocity filter was employed. The emission spectroscopy of the sooting plasma was done during the different stages of the regenerative processes within the source.

The mass spectra contain a broad range of charged clusters and cages from C_2 to $\geq C_{1000}$. The state of the carbon vapor within the source is evaluated by using the characteristic line emissions from the carbonaceous discharge whose formative mechanisms depend upon the kinetic and potential sputtering of the sooted cathode. The discharge initiated with the noble gases, sputters atomic and ionic C and its clusters C_x ($x \geq 2$), noble gas metastable atoms and ions, energetic electrons and photons in the cavity of the graphite hollow cathode. The parameters of soot formation and its recycling depend critically on the discharge parameters, the geometry of the hollow cathode, the graphite hollow anode and the 3D profile of the cusp magnetic field contours.

The C clusters formed in the noble gas discharge gradually transform into the carbon vapor show dependence on the shape and form of the cusp field, in the graphite hollow cathode discharge. The constitution of the carbonaceous discharge characterized and diagnosed by ExB mass spectrometry contains the carbon clusters $C_x^+ \geq 2$ emitted from the source. The state of the carbon vapour within the source is simultaneously, monitored by photoemission spectroscopy at various stages of soot formation and its recycling. The regenerative, sooted source provides an ideal clustering environment conducive to the formation of a wide range of clusters C_x^+ ($x \leq 10^4$) including the linear chains, rings, closed cages and even C onions [35-37]. The mass spectrometry identified the existence of sooting layers on graphite cathode's inner walls as a precondition for the formation of clusters. During these experiments, the photon emission spectra of the atoms, molecules and ions (both positive as well as negative) were obtained from the energetic heavy ion irradiation of the non-regenerative, flat graphite surfaces [36]. The velocity spectra of the sputtered clusters emitted from the regenerative sooting source showed similarities with the results obtained by other workers from the sublimation [12] or sputtering by ion irradiation of graphite [13]. However, the formation of sputtered clusters in a regenerative sooting environment has significant differences. The regenerative sooting source is identified by the state of the excitation and ionization of the carbonaceous discharge. The indicators are the relative number densities of the excited and ionized atoms of the noble gas and carbon. The manipulation of the sooting environments had been shown to be crucial to the formation and discovery of fullerenes in the laser ablated graphite plumes by the Rice university group [2]. Similarly, the essential parameters for the technique of mass production of fullerenes depended on the manipulation of the parameters like the support gas pressure for the dc arc discharges between graphite electrodes [5]. The graphite hollow cathode's noble gas discharges are characterized by the recycling of the cathode deposited soot. The evolving stages of the regenerative sequences build a picture of the carbon cluster formation mechanisms while at the same time, reveal the respective roles played by the neutral, excited and ionic states of C_1, C_2 and C_3 and higher clusters at various stages of the regenerative, carbonaceous discharges.

A compact $E \times B$ velocity filter was specially designed and employed for the mass spectrometry of the carbon clusters [35]. On the basis of the analyses of the velocity/mass spectra, the transition from a pure sputtering mode to a sooting one can be identified. Emission spectroscopy of the excited and ionized components of the discharge and the role played by the excited species in the recycling and regeneration of the cathode deposited clusters, can be explored by the discharge parameters like the discharge voltage V_{dis}, discharge current i_{dis} and the support gas pressure P_g. Kratschmer et al [5] had also identified the gas pressure as a critical parameter for sooting in the dc arc discharge between graphite electrodes. The regenerative sooting source explores different modes of excitation and ionization of the atomic and ionic species by the electrons, the estimation of the excitation temperatures T_{exc} for various discharge species and the conspicuous transition from the C_3 dominated discharge to a sooting plasma containing large clusters and cages.

Unique cluster forming properties of the regenerative sooting discharge were associated with the cusp magnetic field contours [35-37]. In the initial stages of the evolving carbonaceous discharge, kinetic sputtering emerges as the main contributor. The second mode of the discharge can be classified as the potential sputtering mode associated with the high pressure discharges where the density of the ionized species C^+ and Ne^+ considerably reduces. A constant but gentle surface erosion by potential sputtering dominates this mode. Kinetic sputtering also takes place and the process of cathode sputtering involves the two mechanisms (kinetic and potential sputtering) together. Neutral (C_1^*) and ionized carbon (C^+) are the integral constituents of all sooting processes and their relative densities become the indicators of the soot formation on cathode walls. During these sooting modes, agglomeration of carbon clusters on the cathode surface gets recycled or regenerated by the kinetic sputtering with energies up to 500 eV and the potential sputtering through collisions of the metastable Ne* atoms with energies $E \sim 16.7$ eV. The transition from the sputtering proficient regime dominated by the stable C molecule C_3 to the large clusters and cages containing soot is identified and presented later, in this chapter. This information provides us with an understanding of the mechanisms that are responsible for the evolving stages of the formation and fragmentation of the clusters and fullerenes in the regenerative sooting discharges.

2.2. The formation and manipulation of the regenerative soot

The fullerene containing carbon soot was actively studied and reported after the experiments on the laser ablation of graphite followed by the supersonic expansion of the carbonaceous vapour [2]. A large number of researchers dealt with multiple techniques, processes and discussed the mechanisms of the formation of the Buckyball-the C_{60}, as reported in the early books on the science of fullerenes, for example ref. [6]. The book collected and documented the state of understanding of the fullerenes-the closed cages of carbon up to the year of the announcement of the Nobel prize-1996. The agglomerates of the pure carbon clusters so formed included the open and the closed cages. Together these clusters produce what can be regarded and referred to as the *soot*. Various other techniques have been developed for the production of the soot. The Nilore group developed experimental devices and focused on the regenerative soot, investigated and reported different techniques during 1997 and 2002 [35-37]. Transmission electron microscopy groups conducted electron microscopy of the soot with high energy electrons and showed visible evidence of the emergence of the nanostructures of carbon in the form of mono- and multi-shelled nanotubes and fullerenes [4,38,39]. These diverse experimental techniques demonstrated that the soot can be generated, manipulated and tailored to produce the desired nanostructures of carbon.

In yet another technique, high energy ion irradiation in polymers was shown to induce the clustering of carbon atoms leading to optical blackening, electrical conductivity changes as the dominant, high energy ion-induced chemical effects. These changes involved the nuclear as well as electronic energy transfer from the ion to the carbon atoms in the solid matrix that lead to the ion-induced clustering processes. During these irradiations, the orders of magnitude estimate for the size of graphitic islands or the carbon rich zones ranged from 100–500 Å [40]. Similar experiments with MeV heavy ion sputtering of polymers at Uppsala identified the formation of fullerenes in MeV iodine ion bombarded PVDF targets [41]. In these high energy ion irradiations, the fullerene yield measurements as a function of ion fluence indicated clustering to be dependent on ion-induced chemical changes in the polymer. Another experiment with 130 MeV/amu Dy^{22+} ion bombardment of graphite reported that the synthesis of fullerenes during the chromatography of the irradiated samples showed traces of C_{60} [42].

The first phase of the Nilore experiments had also shown that by using 50-150 keV Ar^+, Kr^+ and Xe^+ beams on amorphous graphite, the whole range of clusters including the closed cages can be produced. The ion energy, the dose and angle of ion incidence were shown as the parameters of the soot formation and regeneration [7-9].

2.3. Soot regeneration in carbonaceous discharges

Understanding of the mechanisms of the carbonaceous discharges lies in the synthesis of common features of the widely different physical methods of producing soot. The aim of the design of a tunable, soot regenerative source is to create a recyclable carbon vapour environment and to study the formative as well as the dissociative stages of carbon clusters. Such a source was conceived, experimented with various designs and reported as a cusp field, hollow cathode, carbon cluster ion source [35]. The schematic diagram of the regenerative sooting source is shown in Fig. 2.1. Its operative features depend upon the sputtering efficiency of the cathode with ions of one of the noble gases and the subsequent soot formation leading to the clusters of carbon atoms. A steady stream of carbon atoms is sputtered into the glow discharge plasma from the soot containing graphite hollow cathode. The key to the ignition and sustenance of the discharge in the gas pressure range $\approx 10^{-1}-10^2$ mbar is a set of the six bar magnets wrapped around the hollow graphite cathode providing an axially extended set of the cusp magnetic field contours. Xe,

Kr, Ar, Ne and He have been used to provide the initial noble gas discharge which transforms into a carbonaceous one as a function of the

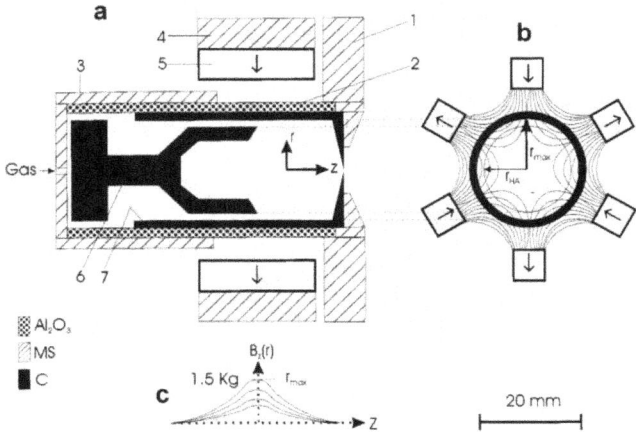

Fig. 2.1. The schematic diagram of the source is shown with graphite Hollow Cathode HC and Hollow Anode HA. The cusp magnetic field $B_z(r,\theta)$ is also shown with arrows. The source components are numbered and shown in (a) with the HC as #6 and HA #7 and a set of six permanent bar magnets indicated as #5. A base plate #1 holds an alumina tube (#2) bonded at both ends with MS rings holding the HC and HA. The set of six bar magnets (#5) are held on the inner surface of an MS ring (#4). (b) shows the r and θ variations of the cusp field lines intersected by HC shown as a thick black circle at $r_{HC} = r_{max}$ and a dotted circle indicates HAs outer radius = r_{HA}. The axial field $B_z(r,\theta)$ is plotted for the six magnets. (c) The cylindrical ring of magnetic steel (#3) shapes the desired $B_z(r,\theta)$ vs. z cusp field contours, enhancing the field in the region between HC and HA. (Fig. from ref. [35]).

discharge conditions. The sooting discharge so produced demonstrates a temporal growth in the densities of sputtered carbon atoms and ions as a function of the discharge voltage V_{dis}, current i_{dis} and the noble gas pressure P_g. The ions anchor onto sets of the field contours, the direction of their consequent gyration motion and clustering probability is determined by collision with electrons, neutral and excited C, and the support gas atoms, and with the walls. The hexapole field confinement is designed so that the radial B_r and axial B_z field lines produce the combined 3D magnetic field contours $B_z(r,\theta)$ shown in Fig. 2.1c. The streams of the gyrating C_1^+, Ne^+ and C_x^+ ions with large collision cross-sections eventually lead to the inside walls of the graphite cathode surface to impact with $E \approx qV_{dis}$, q is the charge on the ion. The ion-impact continuously modifies the graphite cathode surface properties of the sooted substrate. These sooted layers are recycled by kinetic sputtering by the energetic ions as well as potentially sputtered with the metastable species of the discharge [43]. Velocity spectra of the emitted charged species showed dependence of the C cluster emission on the state of the sooting in the source. Spectrometry of the regenerative soot was done by various designs and modifications of the source and the $E \times B$ velocity filter for the detection and diagnostics of large carbon clusters [44].

Different mass analyzing techniques have been employed by researchers for the detection of the large clusters. These include time-of-flight (TOF), momentum and $E \times B$ velocity analysis. Direct recoil spectroscopy was discussed in Chapter 1. For the on-line mass analysis of clusters, the velocity analysis has certain advantages over the corresponding momentum analysis. One such advantage is that the momentum analyzers require increasingly larger diploe magnets (~few T), whereas, velocity filters operating with modest magnetic fields (~0.2-0.3T) can detect the presence of even the heaviest clusters [44]. The comparison of the velocity filtration and the time-of-flight mass detection with the pulsed laser beams, shows that the latter possesses superior resolution especially in the higher mass range. On the other hand, for the in situ investigations of the continuously regenerated soot, the $E \times B$ velocity filter is an appropriate mass diagnostic tool. It will be described in detail in this chapter. In addition, it has demonstrated its utility as a low cost diagnostic tool for the mass detection of the very small to the very large clusters [35,37,44].

2.4 Spectroscopy of the regenerative soot

2.4.1 Mass spectrometry with E × B velocity filter

The permanent magnet based $E \times B$ velocity filter can perform mass analysis in a characteristic way. All masses are deflected by the fixed magnetic field according to their respective masses. The straight through beam contains the desired mass at the compensating electric field $\varepsilon_0 = B_0/v_0$, where B_0 is the magnetic field intensity along the axis and v_0 –the velocity of the particular ion. All of the charged species including monoatomic ions and the ionized clusters are extracted at the same extraction voltage and hence have the same energy; the velocities differ according to their respective masses $v_0 = \sqrt{2E/m}$. A velocity spectrum always contains all masses irrespective of their mass: the resolvability, on the other hand, is dependent upon certain design features discussed below.

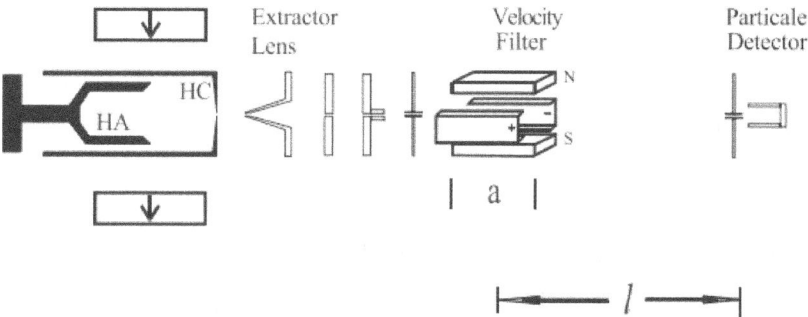

Fig. 2.2. It shows the cluster ion source of Figure 1, extraction lens, collimators, the velocity filter of dimension a. The Faraday cage is at distance l from the filter. The source is composed of a graphite Hollow Cathode (HC), Hollow Anode (HA) and a set of hexapole bar magnets shown with arrows and described in detail in Fig. 2.1.

Figure 2.2 shows the experimental arrangement for the detection of carbon clusters C_x from the regenerative sooting source. It does not show the optical emission spectroscopic arrangement which needed only the spectrometer along the line of extraction or at right angles to it. A well collimated set of extraction lens set up provides a ±0.1° beam to the velocity filter of effective length a. A pico-ampere meter measures the analyzed masses on a Faraday cage l mm away from the exit of the filter, in our case $l = 1500$ mm. Alternatively, a Channel Electron Multiplier was used. The detection of heavy carbon clusters with an $E \times B$ velocity filter depends on the highest possible magnetic field B_0; it sets other parameters accordingly. $B_0 = 0.35$ T on the axis of the filter between the poles that are 10 mm apart. Specially shaped electrodes provide the compensating electric field ε_0 for the straight through masses. These electrodes are slightly extended outwards to compensate for the magnet's edge effects. The resolution of the $E \times B$ velocity filter is determined by the dispersion d of masses $m_0 \pm \delta m_0$ from the resolved mass m_0 that travels straight through the filter with velocity v_0 (= B_0/ε_0). Dispersion $d \propto al(\delta m_0/m_0)(\varepsilon_0/V_{ext})$ where a and l are the lengths of the velocity filter unit and the flight path, respectively. For a given ratio$(\delta m_0/m_0)$, the dispersion d can be enhanced by stacking multiple filter units since $d \propto na$, n being the number of filter units, or increasing the flight path l and also by varying the V_{ext}.

The mass spectrum shown in Fig. 2.3 was obtained from a well-sooted source that was operated for ~100 minutes with Ne gas and $V_{dis}=0.85$ kV and at $V_{ext}=1.0$ kV and $i_{dis}=200$ mA. The spectrum exhibits the range of the linear chains, rings and the closed cages. The smaller cluster peaks of C_1 to C_{11} appear in the larger fraction of the velocity filter's electric field 200-700 V. While the larger clusters are grouped in the smaller, initial range of the electric field i.e., within 0 and 200 Volts. This is a peculiar feature of the velocity selection since the lightest of the masses have the highest of the velocities for a given V_{ext}. Heavier clusters having the smaller velocities are grouped near the origin. Whereas, this limits the resolution of individual heavy masses, these are still detected. The velocity filter has this unique aspect of mass analysis; a given heavy cluster ion is deflected depending on the strength of the magnetic field, while the compensating electric field brings it back to the central path for detection. The filter resolves velocities $v_0 = B_0/\varepsilon_0$ of the straight through ions with variable resolutions. This aspect is evident in Fig. 2.3.

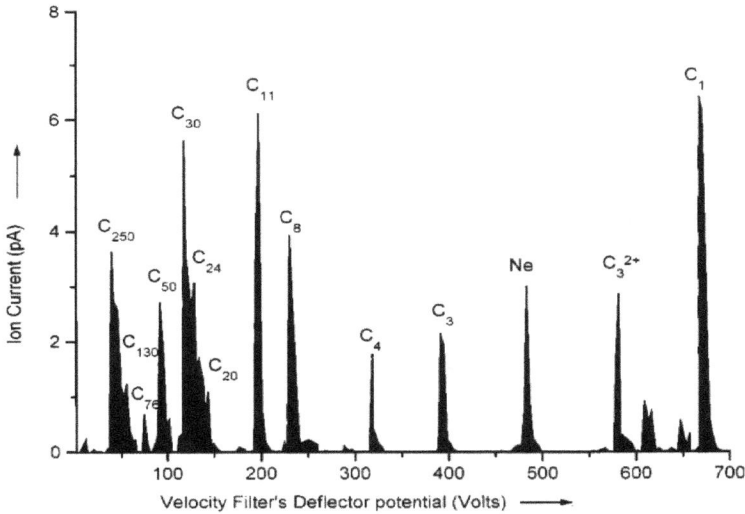

Fig. 2.3. The mass spectrum of the well sooted source with V_{dis}=0.85 kV and at V_{ext}=1.0 kV and i_{dis}=200 mA. The spectrum is composed of the linear chains ($\leq C_{10}$), rings ($10 \leq C_{10} \leq 30$) and the closed cages ~C_{50}-C_{250}.

Figure 2.4 elaborates two distinct stages of the regenerative soot; the first is the initial sputtering dominated stage where lower cluster densities of the predominantly linear chains and rings are obtained from

C_1 to C_{19} as shown in Figure 2.4a. The higher clusters are there but in relatively smaller ratios; the y-axis has ~10 pA as the maximum anion current density for the negatively charged C_4. Fig. 2.4b has the mass

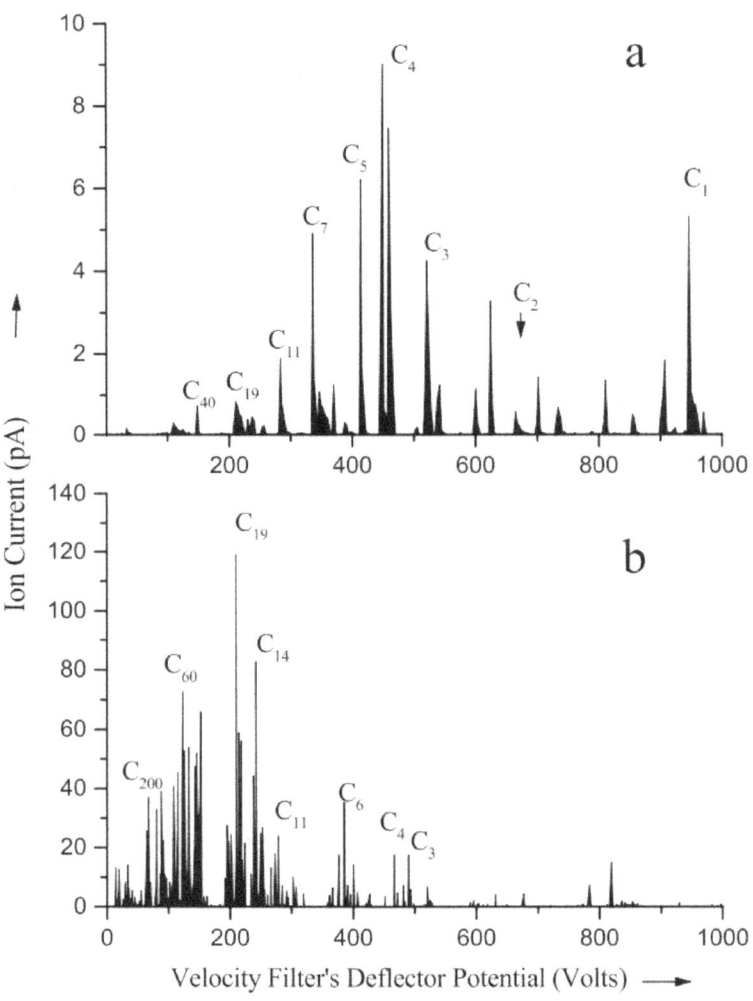

Fig. 2.4. The velocity spectrum of the initial sooting stages of operation with Ne is shown in (a) with pressure in the source P_g =2−3×10^{-3} mbar, V_{dis} =0.5 kV, i_{dis} = 50mA and V_{ext} =2 kV. Clusters from C_1 to C_{19} are present in the spectrum. (b) Shows results from a well sooted source operated with Ne at ≈ 60 watts for 20 hours. It has been obtained with Xe as a source gas. The emitted charged clusters have a range of fullerenes with x ≈ 200 to C_{36}

as well as the rings, chains and linear regimes of clusters. Note the difference in cluster ion intensities in the two spectra. The y-axis shows the current densities of the anions.

spectrum from the later stage of a well sooted discharge obtained with prolonged operation with high power inputs. It produces the entire range of clusters from linear chains and rings to the closed caged fullerenes and perhaps, the carbon onions. Figure 2.4b shows the spectrum of the few peaks identified as C_{60}, C_{200} and the larger ones. It must be noted that the icosahedral C_{60} in the regenerative discharge is not the sole survivor. The soot recycling favors all cages with their associated isomers according to the formation probabilities under the ambient conditions of the regeneration. Anion current densities of the closed caged fullerene peaks, therefore represent the probabilities of formation in the discharge at the time of extraction. The rings and linear clusters are still visible in the heavily sooted discharge, starting from C_{19}, C_{14} to C_{11} and the smaller ones. This is a familiar pattern of the carbon cluster formative and fragmentation environments that has also been observed in the time-of-flight spectrometry of the laser ablated graphite experiments [2,6].

2.4.2 Photon emission spectroscopy

The photoemission spectrum of the characteristic atomic lines and molecular bands of the regenerative soot is shown in Fig. 2.5, recorded with the monochromatic grating blazed at 300nm and minimum resolution of 1 Å [36]. The quantum efficiency of the photomultiplier tube and the relative efficiency of the grating varied between 180–650 nm. Fused silica window fitted on the hollow cathode source allowed the transmission of wavelengths down to ~180 nm. Figure 2.5 presents a typical emission spectrum along with the respective line intensities obtained from the photomultiplier. The peaks were multiplied with the appropriate correction factors for the respective wavelengths while calculating the relative number densities of the excited levels using these spectra. In the figure, the graphite hollow cathode discharge with Ne clearly shows three distinct groups of emission lines between 180 and 650 nm. The first group is between 180–250 nm and it includes emission lines belonging to the neutral, singly and doubly charged C. The CI $\lambda = 1931$Å and $\lambda = 2478$Å are the signature lines emanating from the same singlet level $^1P_1$.

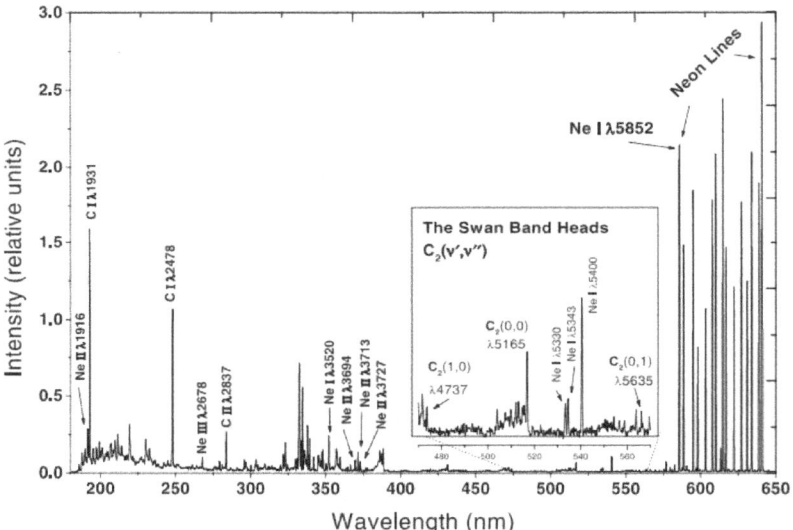

Fig. 2.5. A typical Ne discharge spectrum of the emission lines at $i_{dis} = 200$ mA. The spectrum shows familiar atomic lines due to CI at 1931Å and 2478Å, CII (*i.e.*, C^+ in spectroscopic notation) at 2837Å. C_2's Swan band is shown in the enlarged box with the respective band heads. Excited NeI emission lines have their pronounced presence. The ionized NeII (Ne^+) lines are shown.

The presence of these lines implies that the initially pure Ne discharge has been transformed into a carbonaceous one. Between 300–400 nm, Neon's ionic lines are grouped together with the molecular bands at 357 nm and 387 nm. A significant exception is a NeI line at $\lambda = 3520$Å which is a resonant line of NeI and along with $\lambda = 5854$Å, it can be used for the determination of the excitation temperature T_{exc} of the discharge. The third distinct and high intensity group of emission lines due to the excited atomic NeI lies between 580–650 nm. A large percentage of the discharge power is concentrated in these excited atoms of Ne that cannot de-excite by photoemission to ground. These excited atoms have to give up their energy ~16.7 eV in collisions with the discharge constituents and the sooted cathode walls. We have recently explored their soot regenerative properties as a potential sputtering agent [43]. The emission lines and levels have been interpreted by using NIST's extensive Atomic Spectra Database (ADS) available on the web [45]. There is the clear evidence of the presence of the molecular emission bands due to C_2. The most intense peak is due to the vibrational band head $C_2(0, 0)$ at $\lambda 5165\text{Å}$. The inset of Fig. 2.5 shows the three familiar vibrational band heads found in the atmospheres of carbon stars, in the Cometary spectra and extensively studied using flame spectroscopy where especially, the green head of $C_2(0, 0)$ at $\lambda 5165\text{Å}$ is a signature line of C_2 molecules presence in all carbonaceous environments [49]. Other band heads including $C_2(0,1)$ at $\lambda 5635\text{Å}$ and $C_2(1,0)$ at $\lambda 4737\text{Å}$ have been seen in these spectra. Swan band is the signature of the regenerative soot.

2.5 Mechanisms of regeneration of the soot

2.5.1 Potential and kinetic sputtering of the sooted cathode

The two mechanisms of regeneration of the sooting discharges have been described earlier as the kinetic and potential sputtering of C clusters [43]. The relative contributions of the two processes determine the state of the carbon vapour in all regenerative sooting discharges. Both mechanisms play their respective roles from the initial sputtering of the graphite cathode to the regeneration of the cluster-containing soot. Kinetic sputtering initially generates the binary collision cascades and create the conditions for thermal spikes by the incident ions impacting the surface walls that emit the sputtered species into the plasma. To get the order of magnitude estimates, the Monte Carlo simulation SRIM [46] can be utilized for obtaining the typical monatomic, kinetic sputtering yields. For example, the monatomic yield is $k = 0.12 \pm 0.02$ C_1 atoms/Ne^+ with 500 eV energy incident on graphite. On the other hand, the potential sputtering of the sooted surface takes place upon the interaction of the Ne^* (NeI in spectroscopic notation), metastable atoms with the excitation energy~16.7 eV. Similarly, in the later stages, the excited monatomic and diatomic C species $C_1^* (\equiv CI)$ and C_2^* also contribute effectively to the potential sputtering. Either an individual C_1 or a whole cluster C_x ($x \geq 2$) consisting of many atoms adsorbed on the sooted cathode with binding energies $E(C_x) < E(Ne^*, C_1^*, C_2^*)$, can be ejected with the interaction of the excited plasma constituents. The similar pattern of ion-induced clustering was evidenced in DR spectra shown in Figs. 2.3 and 2.4. It can also be observed in the regenerative sooting source. Here, the clusters that are recycled, can further go through the process of disintegration into smaller units or fragments. The experimentally observed ratio is CI/NeII = 4.5 ± 30% in the pressure range 0.1−1 mbar and is relatively insensitive to the variations of the discharge current i_{dis} [43]. The ratio of the excited C_1 and Ne is CI/NeI ~ 10^{-3}. In Figure 2.6, the relative number densities of NeI, NeII and CI are plotted as a function of the discharge current i_{dis} in Fig. 2.6a, as a function of the gas pressure P_{Ne} in Fig. 2.6b. The average relative number densities of NeII/NeI is ~ $1.3 \times \pm 0.3 \times 10^{-3}$. The carbonaceous character of the sooting discharge is determined by the ratio CI/NeI~ $0.6 \pm 0 : 1 \times 10^{-2}$. The ratio CII/CI ~ $3.5 \pm 0.6 \times 10^4$! It identifies that higher energy electrons have to be present to ionize and further excite these ions. Table 2.1 highlights the significance of the high energy electrons.

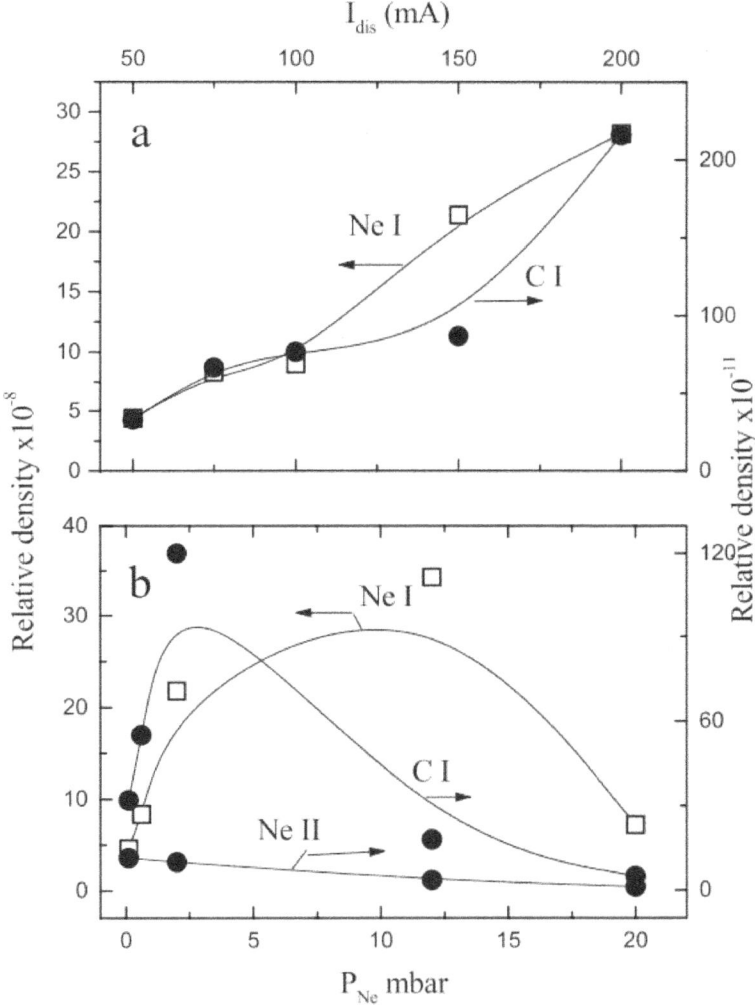

Fig. 2.6. The relative densities of the excited NeI, CI, and ionized NeII versus the discharge current i_{dis} in (a) and P_{Ne} in the range 0.6 to 20 mbar in (b).

The sputtering yield data of Monte Carlo simulations [46] for the ratio C$_1$/NeII can be used to get $k \sim 0.12$ C$_1$/Ne$^+$. The kinetic sputtering provides $\sim 10^{-5}$ CI excited atoms in the $^1P_1$ level, which is a much smaller ratio of the observed relative density CI $\approx 4.5 \times$ Ne$^+$. The potential sputtering on the other hand, generates between 0.5 and 1 CI per NeI. This would yield $\sim 10^{-2}$ excited CI atoms for each NeI metastable atom, as the sputtering is occurring from a sooted cathode that is covered with many monolayers of a loose agglomeration of clusters. These clusters contain on the average 50–100 C_1 atoms [35]. The potential sputtering of the clusters estimates is within 50% of the observed value. The high number density of CI that is observed is primarily due to the potential sputtering of the carbon clusters adsorbed on the sooted surface by the metastable atoms and clusters. The potential sputtering is the dominant soot regeneration mechanism in the graphite hollow cathode carbonaceous discharges that are started and sustained with the noble gases.

Table 2.1. Ionisation rate coefficient α_i cm^3 s^{-1} for the atomic and ionic species of C and Ne, CI-CII, CII-CIII, CIII-CIV and NeI-NeII, NeII-NeIII. All ionisations are from the ground state.

	1 eV	10 eV	100 eV
CI–CII	1.93×10^{-13}	1.7×10^{-8}	9.7×10^{-8}
CII–CIII	1.2×10^{-19}	1.3×10^{-9}	2.6×10^{-8}
CIII–CIV	4×10^{-30}	5.8×10^{-11}	7.4×10^{-9}
NeI–NeII	1.15×10^{-18}	1.32×10^{-9}	4.2×10^{-9}
NeII–NeIII	3.25×10^{-28}	9×10^{-11}	1.8×10^{-8}

2.5.2 The two energy regimes of electrons in hollow cathode glow discharges

The levels of excitation and ionization of the support gases and the sputtered species cannot be explained by a single electron energy regime *i.e.*, the excitation temperature T_{exc} of any two levels of a species. It is well documented that the hollow cathode glow discharges are initiated and sustained by two well defined electron energy regimes [47]. In For the regenerative soot, the higher regime has $E_e \geq 10$ eV while the other electron energy range can be evaluated from the excitation temperature obtained from the emission lines of NeI, NeII and, also from C_1^* and C_2^*. These provide us an average kinetic energy of electrons $\langle E_e \rangle \approx T_{exc} \leq 1\ eV$. The role played by these two distinct energy regimes of electrons provides an explanation of the rather high densities of the ionized species CII, CIII, NeII and NeIII that are present even at i_{dis} as low as 50 mA. Table 2.1 is prepared by using Lotz's' semi empirical formulation [48] for the ionization rate coefficients α_i cm^3 s^{-1} for the successively higher ionization stages of C and Ne. For these calculations Maxwellian velocity distribution for the electrons is assumed and all excitations and ionizations are from the ground state. At $T_e \approx 1$ eV or less which can be approximated as the discharge temperature in the regenerative soot, the presence of the higher ionized species is much less probable. However, at higher electron energies $E_e \geq 10$ eV, a significant increase in α_i occurs. Between 10–100 eV energy range, the electrons can ionize all ionic stages of C and Ne with similar orders of magnitude probabilities. The spectral line ratios of the C ions have been used in Table 2.1, to evaluate T_e for the carbon impurities in the tokomak plasmas for CII to CIV in the $T_e \sim 4 - 40$ eV range. Since in our case the $T_e \sim 1$ eV, the required high energy electrons are provided by the cathode for the ionization for the higher ionization levels of C and Ne. These are available due to the secondary electron emission from the cathode but in much reduced intensities compared with the thermal electrons.

2.5.3 The state of the atomic and ionized C

Figure 2.7 shows two spectra with the source operating at gas pressure $P_{Ne} \approx 0.6\ mbar$. During the first spectrum in Figure 2.7a with i_{dis} = 200 mA, the discharge voltage V_{dis} = 0.6 kV. This spectrum is taken on a freshly prepared cathode surface, at the glow discharge starting values of V_{dis} and i_{dis}. This is a typical carbon cathode sputtering spectrum. The dominant Ne emission lines, atomic and ionic are present.

The VUV part of the spectrum is dominated by the neutral, singly and doubly charged CI, CII and CIII. In Figure 2.7b the emission lines λ = 1931 Å and λ = 2478 Å are the two dominant VUV lines of CI that can be seen along with the λ = 2135 Å and λ = 2191 Å of CII and the λ = 2297 Å of CIII. Also present is CII's 233 nm intercombination multiplet. It has very small transition probabilities ~0.1s^{-1} for all five lines of the multiplet. This multiplet is generally a weak intersystem transition route for the de-excitation of CII in carbon plasmas.

From the energy level scheme of C in NIST's Database [45], out of the total of 254 CII emission lines between 0–2000 Å, 73 lines are emitted by de-excitation to the first excited level $2s2p^2\ ^4P_{1/2,3/2,5/2}$ of CII. This level, in turn de-excites to the ground level $^2P_{1/2,3/2}$ by the emission of the 232 nm multiplet. The intense emission indicates that CII exists as a highly excited C ion in the discharge. The NeI lines between 580–650 nm remain as the significant emission feature of all of these spectra.

Singly ionized carbon's first excited state $^4P_{1/2,3/2,5/2}$ has lifetime τ ~4.7 ms. Thus its level density serves as a useful indicator of the carbon content of the cusp field, graphite hollow cathode plasma. The calculated relative densities from the line intensities of CI λ = 1931Å, CII λ = 2324–2328Å, NeI λ = 5852Å, NeII λ = 3713Å provide us with the statistical data of the carbonaceous discharges.

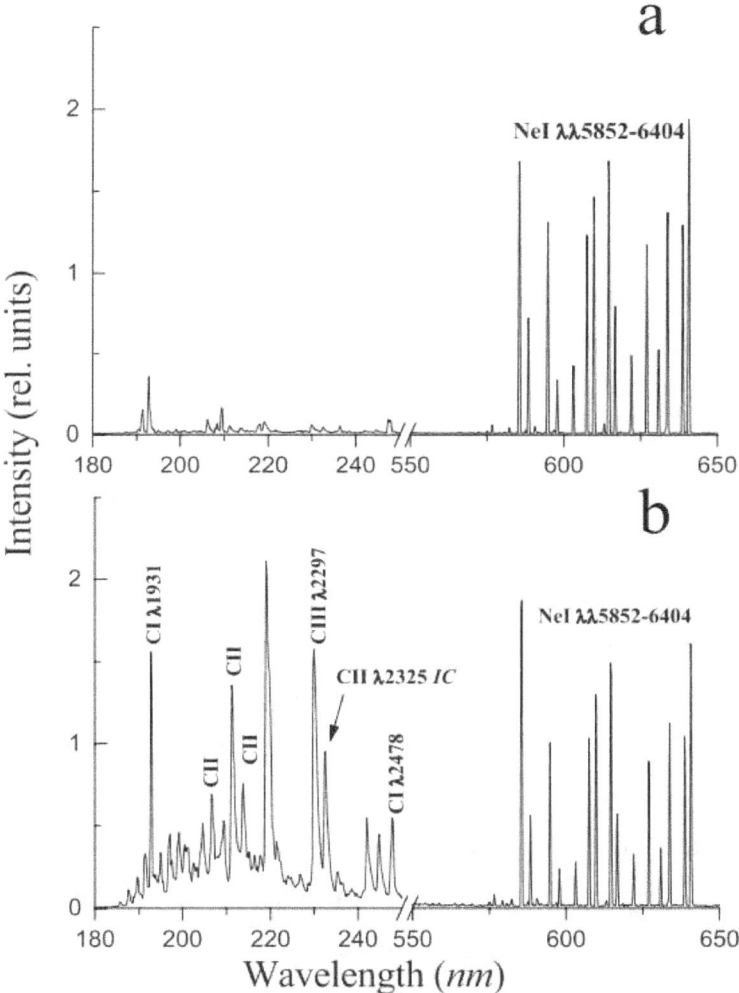

Fig. 2.7. Photon emission spectra are presented with Ne as the support gas at two different discharge voltages with $P_g \approx 0.6$ mbar, $i_{dis} = 200$ mA being kept constant. For (a) $V_{dis} = 0.6$ kV and (b) $V_{dis} = 1.5$ kV. The relative lines intensities are plotted against the wavelength. The x-axis is broken in the two wavelength ranges of $180-250$ nm and $550-650$ nm, respectively.

From the natural radiative lifetimes of these four excited states, the CII has six orders of magnitude longer residence time in the plasma. This has already been discussed earlier that collisions of CII with the walls are more effective as opposed to the short lived constituents. In the i_{dis} range of 50–200 mA, the ratio of the densities $D_{CII}/D_{CI} \sim (3.5\pm0.5)\times10^4$. Similarly, $D_{NeII}/D_{NeI} \sim (1.3\pm0.3)\times10^{-3}$ and $D_{CI}/D_{NeI} \sim (0.6\pm0.1)\times10^{-2}$. These results identify a carbonaceous plasma with the ionized C whose ratio with the excited Ne is $D_{CII}/D_{NeI} \sim 2\times10^2$. These results also imply that the density of the singly ionized neon D_{NeII} is only $\sim 10^{-5} \times D_{CII}$ throughout the above mentioned discharge current range. Therefore, the carbonaceous discharge is dominated by the ionized C and has a 2–4% excited Ne.

2.6 The C_1, C_2 and C_3 content of the regenerative soot

C_2 and C_3 are the essential ingredient of the sputtered species from graphite, c nanotubes as well as from the sooted discharges. But the relative ratios of their neutral and charged states show large variations. These variations depend on the underlying mechanisms that are operating. Honig [12] had mass analyzed the subliming clusters from a graphite oven and measured the ratio of the relative densities as $C_1:C_2:C_3 = 1:0.37:2.83$, While for the negative species, the data indicated much larger share for the diatoms as $C_1^-:C_2^-:C_3^- = 1:3600:40$. In the experimental results the velocity spectra of the positively charged carbon clusters $C_x^+ (x \geq 2)$ from the regenerative sooting discharge is dominated by clusters with large carbon content $C_x^+ (x \geq 4)$. C_2 is present but only as a minor constituent. On the other hand, the positively charged clusters C_x^+ that are kinetically sputtered from a flat graphite disc, are likely to auto neutralize. That may be the reason that in a non-regenerative environment, only the negative clusters C_2^- along with C_3^- and C_4^- were observed in the mass spectra of the sputtered graphite species [35,36].

A typical emission spectrum from the regenerative sooting discharges was shown in Fig. 2.5. In similar experiments, the Swan band heads of C_2: $C_2(0,0)$ at $\lambda = 5165$Å, $C_2(1,0)$ at $\lambda = 4737$Å, $C_2(2,1)$ at $\lambda = 4715$Å, and $C_2(0,1)$ at $\lambda = 6535$Å were observed. The Swan band

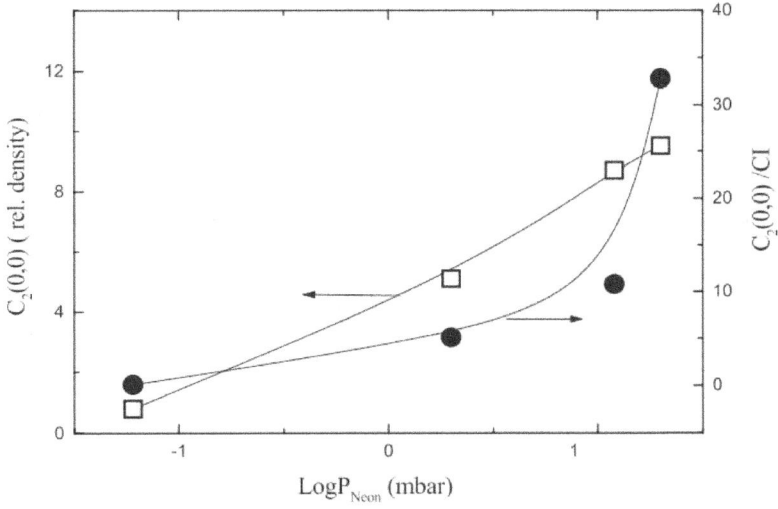

Fig. 2.8. The emission spectra yield the number densities of the vibrational excited $C_2(0; 0)$ at $\lambda = 5165$ Å and its ratio with the excited C atoms CI at $\lambda = 2478$ Å is plotted as a function of P_{Ne} between 0.06 and 20 mbar at $i_{dis} = 75$ mA. The $C_2(0; 0)$/CI ratio rises steeply after $P_{Ne} > 1$ mbar.

of the C_2 is a useful indicator of the existence of C_2 in the sooting discharges. An increase in the number density of neutral C_2 as a function of the discharge current i_{dis} follows a similar pattern of direct proportionality to i_{dis} by the atomic species C_1. Whereas, on increasing the support gas pressure from 0.06 to 20 mbar at constant i_{dis} as shown in Fig. 2.8, there is an inverse relationship between the number densities of the atomic and diatomic carbon species. The derived number densities of the excited CI and C_2 from the line intensities of the electronically excited CI at $\lambda = 2478$ Å and the lowest transition of the Swan band with the vibrationally excited $C_2(0,0)$ at $\lambda = 5165$ Å. $C_2(0,0)$ number densities, are plotted on the left vertical axis and the ratio $C_2(0,0)/CI$ along the right vertical axis. $C_2(0,0)$ is an increasing function of both of the discharge parameters i_{dis} and P_{Ne}. CI also increases rapidly with i_{dis}, but its contribution is reduced at higher pressures as can be seen in Fig. 2.7b. From the photon emission data, the ratio of the singly charged to the excited C *i.e.* CII/CI remain constant, for example in the range of P_{Ne} = 0.1−1 mbar it is 0.55±0.2 for i_{dis} = 50 to 200 mA. Similarly, $C_2(0,0)/CI$ = 2.2 ± 0.4 under the same conditions. At low pressure discharge *i.e.*, $P_{Ne} \leq$ 1 mbar the charged atomic carbon CII and the excited diatomic molecular carbon ($C_2(0,0)$) are directly related with CI *i.e.*, C_1 in the $^1P_1$ level ($E1_{P1} \approx$ 7.5 eV).

2.7. The transition from the C_3 dominated discharge to the sooting plasma

The mass spectra of the positively charged clusters emitted from the source operated at low pressures with a well confined discharge has certain unique characteristics. It is dominated by C_3^+ and much smaller densities of C_1^+ and C_2^+ [12,50]. Two such spectra are shown in Fig. 2.9 by confining the discharge within the annular region between the hollow cathode and the extended hollow anode. The physical confinement between the thin annular region is supplemented by the cusp magnetic field $B_z(r, \theta)$ contours (see Fig. 2.1c). The two spectra shown in Figure 2.9 were obtained with i_{dis} = 150 mA (Fig. 29a) and i_{dis}= 12.5 mA (Fig.

2.9b). The C_3^+ is the major surviving species and at low i_{dis}, in addition, it is the discharge's main ionic component. However, in the photoemission spectra from at these conditions *i.e.* i_{dis} = 12.5 mA, the atomic and ionic C lines (CI, CII, C$_2$ etc.) are present. The velocity

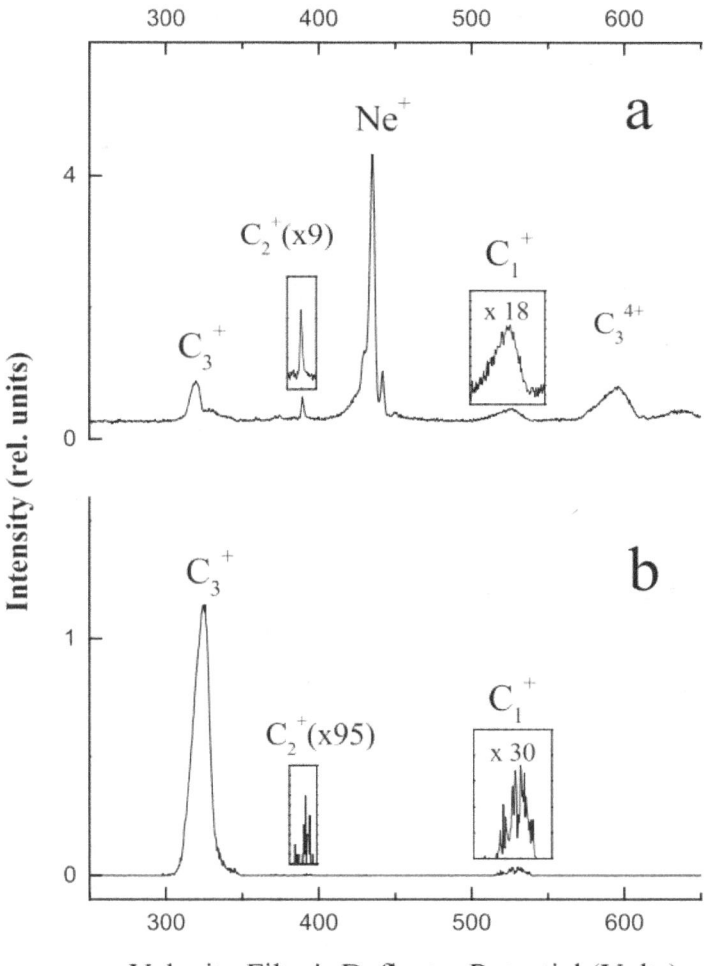

Fig. 2.9. (a) i_{dis} = 150 mA, Ne$^+$ is the most significant peak followed by C_3^+ and C_3^{4+}. Inset: C_2^+ and C_1^+ are shown magnified by a factor of 9 and 18, respectively. (b)) i_{dis} = 12.5 mA, C_3^+ is the only and most significant peak. Inset: C_2^+ and C_1^+ are shown by their respective enlargement factors.

spectrum at i_{dis} = 150 mA shows C_3^+ and Ne$^+$ as the main discharge features. Reducing i_{dis} by a factor of 20 to i_{dis} = 12.5 mA, the C_3^+ is the only surviving and ionized species. In Figure 8 the peaks due to C_2^+ are enlarged by factors of 9 and 95, respectively. Whereas, the C_1^+ intensity is seen to be enlarged by 18 and 30 times in the two respective figures. The broad peak due to C_3^{4+} is also present in the spectra.

The accumulated data for the relative ion densities of C_3^+, C_2^+ and C_1^+ as a function of i_{dis} and P_{Ne} are presented in Fig. 2.10. The ratios C_3^+/C_1^+ and C_3^+/C_2^+ are plotted as a function of i_{dis} from 12.5 to 150 mA. C_3^+ is the sole survivor at very low discharge current. Its relative number density increases with respect to that of C_1^+ by a factor of 27 for decreasing i_{dis} = 150 mA to 12.5 mA. Whereas, C_3^+/C_2^+ increases by two orders of magnitude as i_{dis} is decreased in the same range. The number density of C_2^+ increases from 5% to 20% of C_1^+. From the tabulated data presented in Figure 2.9 we conclude that C_3^+ is not only the significant species at low i_{dis} but is the main constituent of the discharge under the experimental conditions. The predominance of C$_3$ in such a discharge may be due to the regeneration pattern of all clusters C_x^+ ($x > 3$) up to C$_{30}$ favors the accumulation of C$_3$ as the end product. This fragmentation scheme $C_x \rightarrow C_{x-3} + C_3$ with dissociation energy $E_{diss} \approx 5.5 \pm 0.5$ eV has been predicted and experimentally verified for all C_x^+ up to $x \leq 60$ [51-53]. Their end product C$_3$ can itself dissociate into C$_2$ and C$_1$ via C$_3 \rightarrow$ C$_2$ + C$_1$ with $E_{diss} \approx 7 \pm 0.5$ eV. This fragmentation pattern may explain the preponderance of C$_3$ as well as the enhanced contribution of C$_1$ at higher i_{dis}. At higher source pressures P_{Ne}, the large relative increase in the density of C$_2$ cannot be explained by the C$_3$ fragmentation alone. For example, at $P_{Ne} > 1$ mbar, the increased contribution from the C$_2$(0, 0) is accompanied by a consequent decrease in the excited and ionic C1 (CI, CII, CIII) lines. This was interpreted it as the onset of the formation of the closed caged clusters C_x ($x > 30$). These clusters further fragment via C_2 emission $C_x \rightarrow C_{x-2} + C_2$ ($x \geq 30$). This is the enhanced $C_2(0, 0)$ intensities at $P_{Ne} > 10$ mbar are observed [50].

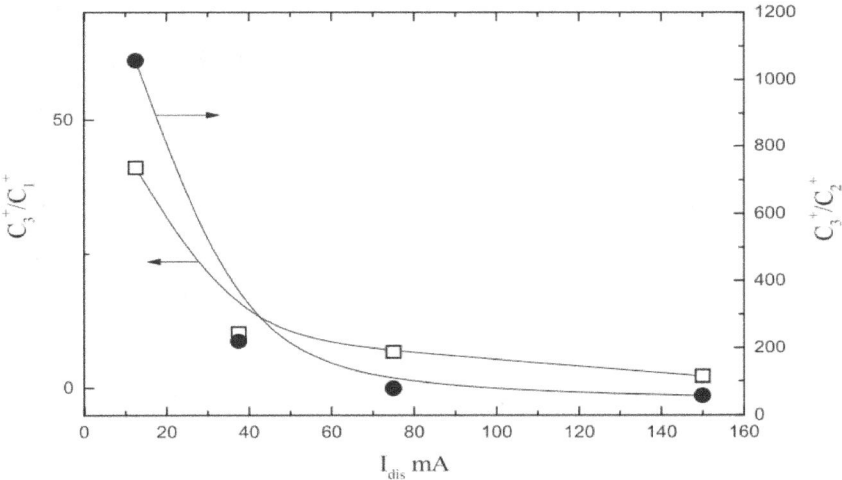

Fig. 2.10. The ratios of the ion densities C_3^+/C_1^+ and C_3^+/C_2^+ are plotted as a function of i_{dis}. C_3^+ is the main discharge constituent and its contribution increases by lowering the i_{dis}.

2.8. The Summary of the Regenerative Soot

The simultaneous mass spectrometry of the clusters extracted from the source and the photon emission spectroscopy of the carbonaceous discharge yields the information about the formation, dissociation and fragmentation of the clusters within the sooting discharges. Collectively, these sequences constitute the formative and fragmentation stages of the regenerative soot. The main agents for the regeneration of the soot were identified as a) the kinetic and potential sputtering from the sooted cathode, b) the collisions of the soot constituents with the two different regimes of energetic electrons, and c) the collisions between the metastable, excited and ionized $Ne^{*,\pm}$ and $C_x^{*,\pm}$.

The inclusion of carbon into the neon plasma is identified from the data on the relative number densities of the excited atomic species CI and NeI calculated from their characteristic lines' intensities in the emission spectra. The initiation of the discharge at a given gas pressure requires high T_{exc} regimes with higher values of V_{dis} and i_{dis}. The sputtering of graphite cathode is followed by the excitation and ionisation of these sputtered C atoms with the high energy electrons that are emitted from the cathode and accelerated in the cathode fall to energies ~ 500 eV. This sequence forms the main agent for the discharge sustenance with the gradually increasing C content. The singly ionised CII content participates efficiently in the kinetic sputtering of the cathode along with NeII. Due to the presence of the large fraction of the negative species, the possibility of $C_x^+ + C_y^- \rightarrow C_{x+y}^{0,\pm}$ interactions may be a contributing step towards larger cluster formation. It was shown that the soot formation in the hollow cathode discharge may proceed in two distinct stages:

(1) The sputtering dominated regime with the discharge produced and contained within the annular region between the cylindrical graphite anode and cathode. C_3 dominated the non-LTE discharge as the sole survivor of the linear chains and ring type clusters C_x ($x \leq 30$). C_3 itself emerges as the end-product of the fragmenting C_x in the discharge. Stability of C_3 has been the subject of investigations as a stable C molecule [51-54].

(2) The sooting discharge emerges if it is allowed to expand into the

extended graphite hollow cathode, one obtains the soot formative environment. Larger clusters C_x ($x \geq 30$) are formed leading to the cage closure resulting in the formation of fullerenes. The cage closure amounts to carbon accretion at a very rapid rate during the sooting stages as opposed to the sputtering one described in (1). These experimental observations will be employed in the development of the theoretical models to describe the emergence of the closed cages from the self-organizing regenerative soot in the next Chapters.

Note: Figures 2.3 to 2.10 are from ref. [37].

The Self-Organizing Soot

Chapter 3

A Continuum Elastic Model of Nano Curvature

3.1. Introduction

The spherical curvature induced by pentagons in the corannulenes and the hexagonal sheets can be shown as the basic constituent that controls the growth of fullerenes and single-walled carbon nanotubes in the regenerative soot forming environments. Formation of the initial ring of five or six atoms is the essential step which, with the addition of further pentagons and hexagons, determines whether a spinning fullerene is to be formed or the cap that lifts up and leads to the formation of a SWCNT. In this chapter, the continuum elastic arguments are used to determine the criteria for the growth of these structures in the soot forming, carbonaceous environments by utilizing a model of the self-organizing carbon cages [55].

The experimental evidence for the growth of closed cages of carbon that included the spherical fullerenes, has been presented and discussed in Chapters 1 and 2. The present chapter deals with the mechanisms by which the cage closure occurs leading to the fullerenes and also the formation of nanotubes in the experimental conditions of dc arc discharge, laser ablation, CVD and the regenerative soot source, where anyone or the both types of the carbon nanostructures can be produced. The objective is to understand and explain the nature of the spherical curvature by extending the theory of continuum elasticity to nanostructures. Variations in the experimental parameters like the electrode geometry, gas pressure, and temperature of the substrates and the presence of catalysts like Fe, Co, Ni have generally quoted as being the reasons for the preferential growth of the fullerenes or nanotubes [6]. In this chapter, the criteria are derived, using the continuum elastic arguments, to identify the mechanisms that produce either the fullerenes, the nanotubes or the both types of nanostructures, in a coherent manner. The corannulene road for the growth of fullerenes has been proposed using topological arguments [56]. In this chapter, the physics and the mechanisms of the incorporation of the curvature inherent in the pentagon-centered corannulene, are described in the nanostructure-applicable, continuum elastic model, where the cage closure is explained and the criteria that govern the growth of fullerenes and SWCNTs in sooting environments are deduced.

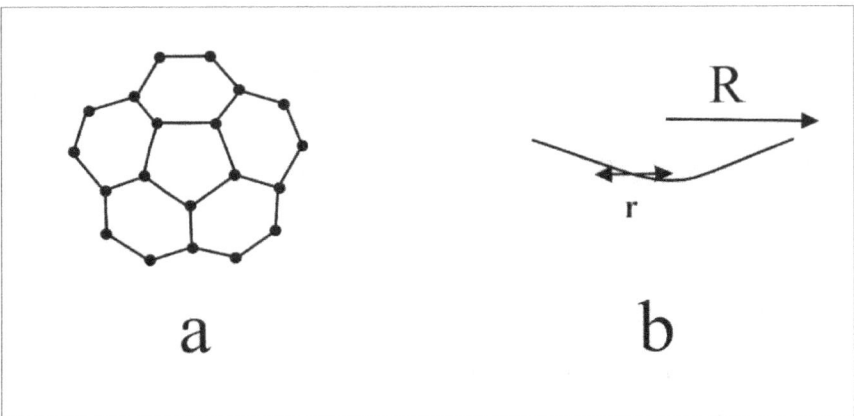

Fig. 3.1. The corannulene bowl of 20 C atoms arranged as the 5 hexagons around a central pentagon (a) is shown with 20 C atoms at the vertices. (b) The side view of the bowl is shown as the protrusion centered around a pentagon with r~1-2Å in the larger curved sheet of R~3-4Å consisting of the hexagons only.

The continuum elastic model for the shelled structures of carbon demonstrates that the curvature-related growth of the fullerenes and nanotubes can be associated with the formation of the initial bowl of corannulene or by the addition of pentagons in an all-hexagon sheet. Corannulene in Fig. 3.1. is shown to be, one of the basic structural unit in the self-organization of carbon nanostructures. It will be shown, later in the chapter that the fullerenes can grow in both situations, in which either a pentagon or hexagon is the first stable ring to be formed. The model describes the topological and continuum elastic conditions of the C_{60} to be the most dominant species among the fullerenes produced under optimum conditions in carbonaceous discharges. The conditions of growth of a SWCNT around a hemispherical cap of C_{240} in the armchair configuration will be demonstrated as the preferential single-walled nanotube with diameter ~14Å. The SWCNTs grown in a catalyst-free environment by laser ablation and the dc arc discharge between C electrodes also show an optimized nanotube diameter of 14Å [57-59]. The continuum elastic considerations of the curvature introduced by the six corannulenes that are necessary for the cap of the SWCNTs, emerges as the condition of the non-abutting corannulenes for the dominant growth of the 14Å diameter SWCNTs. A similar condition of the non-abutting pentagons has been proposed and verified for the emergence of the icosahedral I_h isomer of C_{60} out of the 1812 isomers [3]. As a topological corollary, the non-abutting corannulenes is proposed here, as a growth requirement of the SWCNTs.

3.2. Continuum elastic model of the closed cages of carbon

Various researchers have studied the nano-spheroidal C cages, and the C nanotubes capped with half-fullerenes, by utilizing the continuum elasticity theory [60-69]. In the present Chapter, the curvature is explained using the **stretching** and **bending** aspects for describing the curvature induced in the shells [70]. The spherical curvature inherent in the nascent sp² bonded corannulene, or the pentagon-induced curvature in graphene sheets emerges as the fundamental growth requirement for closure as the local **stretching** of the shells of C nanoparticles. This approach is characteristically different from that based on the **bending** of the graphene sheets to produce spherical curvature [60-66].

In developing the nano-elastic model of C-shells it is clarified that fullerenes or their respective half-caps are not comparable to the bent graphene sheets. This is because the elastic behavior of shells versus plates is manifested in the **stretching** and **bending** effects which appear in reverse order for the two kinds of structures [70,71].

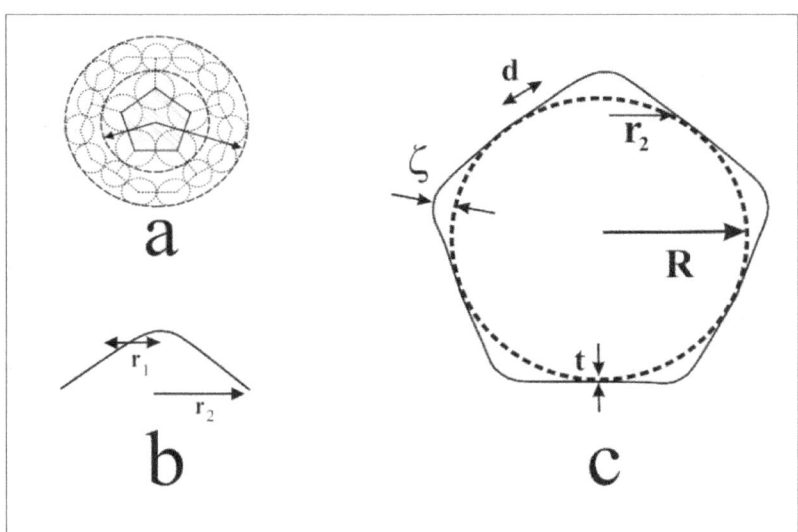

Fig. 3.2. (a) The pentagon-centered corannulene is shown with two effective dimensions shown by arrows. (b) The same 20-atom structure (1 pentagon + 5 hexagons) of the corannulene is shown tangentially to reveal the role that r_1 and r_2 play in the formation of shells. (c) A typical, complete icosahedral shell is shown with the outward protruding corannulenes. The outward protruding structure is superimposed on an inner sphere of radius R, thickness t; the protrusion is shown to have dimension $\sim\zeta$ as the difference between the radii of the circumscribing and inscribing spheres. The corannulene caps $\sim r_2$

are along the meridian with the dimension **d** $\sim \sqrt{tR}$. (From ref. [55]).

Existence of the spherically curved surfaces induces additional forces that depend on the properties of the particular surfaces. The coefficient of surface tension T is related, in the case of the separation of phases, to the pressure difference ΔP across the interface by Laplace's equation $\Delta P = 2T/R$, where R is the radius of curvature. Elasticity theory relates T to the tangential stresses. In the case of a C nano-shell, the equilibrium equation requires the calculation of these stress tensors. Any localized deviations from the global sphericity of the cage are dealt with by taking **stretching** as the first-order and **bending** as the second-order effect. On the other hand, the conversion of a 2D graphene sheet into a cylindrical tubule does not require the stretching effect to be considered; bending is the first-order effect. The strain energies obtained for tubules with the consideration of bending only can, and do provide fairly accurate estimates [60]. The evaluation of the deformation energies that are required for the pentagonal protrusions in shelled structures requires the **stretching** and **bending** effects to be taken into account together.

The following four assumptions have been made to establish the nanoelastic character of the C shells: (1) The fullerene shell is assumed to be equivalent to an appropriate Goldberg polyhedron with 12 pentagons and a variable number of hexagons [3]. (2) The thickness **t** of the shell is expected to remain constant even during deformations of the shell. Here, **t** is treated as the thickness of a fullerene shell yields an average C atom diameter of 1.82 Å. This value of **t** is used for calculations. (3) The elastic modulus $Y = 10^{12}$ Pa and Poisson ratio $\nu = 0.163$ remain constant in fullerenes and the cap nanotubes of different radii; we use the average numerical values for the basal plane of graphite from [60]. (4) When dealing with the pentagonal deformations, the **stretching** and **bending**

3.3. Elastic properties of the graphene sheets

The internal stresses which occur when a body is deformed locally are due to the forces of interaction between atoms. These have short ranges of the order of inter-atomic distances. Such forces when acting on any part of the body are considered to act locally on the surface. In the theory of elasticity, being a macroscopic theory, generally the distances considered are large compared with inter-atomic separations. However, in fullerenes, by considering these local deformation-induced internal stresses, insight into their elastic stability related properties can be achieved. The analogy of the nanometer scaled C shells with macroscopic hollow structures and shells has prompted various researchers to employ the continuum elasticity theory [55,60-68]. Nano-elasticity of the fullerenes may be considered an attempt to relate the structure related deformations of the closed cages of carbon with the resulting surface stresses. Majority of these authors have used the continuum elasticity theory to describe fullerenes as bent graphene. The model used in this Chapter extends the continuum mechanics to describe various growth related properties of fullerenes and nanotubes by introducing the stretching and bending effects of the spherical structures' deformations. Elastic properties of the spherical C shells including the fullerenes and nanotubes are different from those of the graphene, which may be considered equivalent to thin plates or flat sheets.

Theory of continuum elasticity [71,72] treats the deformation of plates in two distinct regimes; (a) thin plates with small deformations $\zeta \leq t$ where t being the plate thickness, and (b) for the large deformations ($\zeta \gg t$) that occur due to the application of large external forces or as it will be shown, in the case of fullerenes, due to the spherical curvature induced by pentagons. Pure bending is the dominant effect in the former case while both bending and stretching effects are important in the case of the latter. By employing the continuum elasticity of macro-surfaces, plates and shells, the elastic behavior of the spherically bent sp^2−bonded C sheets and spheroidal fullerenes with pentagonal protrusions will be treated in the following sections.

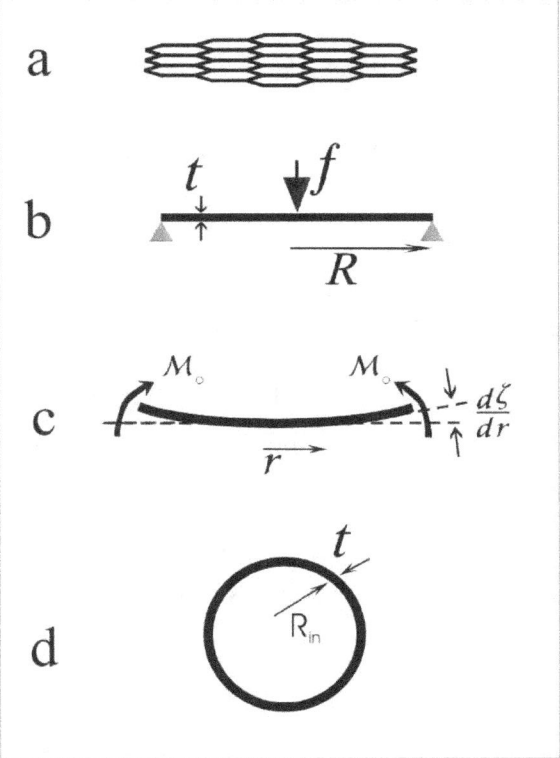

Fig. 3.3. Bending of graphene sheet. The sequence of deformations is shown leading to small deformation in a 19 hexagon, graphene sheet of thickness t and flexural rigidity $D = Yt^3/12(1-v^2)$, where Y is Young's modulus and v the Poisson ratio. In (a) an external applied force f is acting to bend it with the associated bending moments $M_o \approx (1+v)D/r$ at its edges. For circular supports as shown in (c) spherical deformation of curvature $1/r = DM_o/(1+v)$ will occur. For the parallel supports shown in (b) the cylindrical bending of the sheet into an open ended tubule with inner radius R_{in} could result as shown in (d).

In the case of small deformations, ζ denotes the vertical displacement of a point on the neutral surface on which neither extension nor compression takes place. On a neutral surface the displacement vector **u** has components $u_x = u_y \approx 0$ and $u_z = \zeta(x,y)$. In such cases the strain tensor is

$$u_{\alpha\beta} = \frac{1}{2}\left(\frac{\partial u_\alpha}{\partial x_\beta} + \frac{\partial u_\beta}{\partial x_\alpha}\right).$$ The total free energy of a plate for small deformations as a function of ζ is

$$E_{plate} = \tfrac{1}{2} D \iint \left[\left\{\frac{\partial \zeta^2}{\partial x^2} + \frac{\partial \zeta^2}{\partial y^2}\right\}^2 + 2(1-v)\left\{\left(\frac{\partial^2 \zeta}{\partial x \partial y}\right)^2 - \frac{\partial^2 \zeta}{\partial x^2}\frac{\partial^2 \zeta}{\partial y^2}\right\}\right] dxdy \quad \text{eq. (3.1)},$$

where D is the flexural rigidity of the plate $D = Yt^3/12(1-v^2)$. By minimizing this energy one obtains the equation of equilibrium of the plate. Dividing the integral in two parts and varying these separately,

$f = D\Delta^2\zeta$, where the 2D Laplacian $\Delta \equiv \frac{\partial^2}{\partial x^2} + \frac{\partial^2}{\partial y^2}$. This is the equation for the equilibrium of plate bent by the application of the external force f on it. This equation, is solved by setting up appropriate the boundary conditions; a fairly complex procedure. Considerable simplification is achieved if the plate edges are clamped or supported. In the case of the supported plate, for example, shown in Figure 3.3(c), one has $\Delta^2\zeta = 0$ everywhere except at the origin and the result for the force in terms of ζ and plate dimensions is

$$f \approx 16\pi D\zeta / \left[(R^2 - r^2) - 2r^2 \log\frac{R^2}{r^2}\right] \quad \text{eq. (3.2)}.$$

Figure 3.3 shows the sequence of deformations that can lead to small spherical deformation in a graphene sheet with 19 hexagons, of thickness t in Fig. 3.3(a) under an external applied force f in (b) to a plate under bending moments $M_o \approx (1+v)D/r$ at its edges as shown in (c). If the supports shown in Fig. 3.3(b) are not circular and instead are parallel, then the cylindrical bending of the sheet into an open ended tubule with inner radius R_{in} could result in Fig. 3.3(d). The order of magnitude estimate of the pure bending energy E_{plate} in eq. (3.1) can simplify the physical considerations. The first derivatives of ζ are of the order of ζ/R and the second derivatives $\sim \zeta/R^2$. Thus, following ref. [70]

$$E_{plate} \sim Yt^3\zeta/R^2 \qquad \text{eq. (3.3),}$$

this is the order of magnitude estimate for the bending energy per unit area of flat plates under external force f per unit area. It shows a linear relationship of the bending energy with the deformation.

3.4. Large deformations of graphene sheets

Let us first establish the nano-elastic limits to large deformations of graphene sheets (plates of nanometer thickness) to yield the corannulene-like structures that may lead to the formation of fullerenes or the capped nanotubes, and then compare and consider the respective energies of such bent graphene with the spherically curved bowl of corannulene due to a central pentagon. In such a case, the **bending** and **stretching** are involved and there exists no neutral surfaces in large deflections, unlike the situation where $\zeta \leq t$. The infinitesimal strain tensor for the simultaneously bent and stretched plate is expressed as a function of displacement vector **u** for pure stretching and transverse displacement ζ. The 2D infinitesimal strain tensor is $u_{\alpha\beta} = \frac{1}{2}(\frac{\partial u_\alpha}{\partial x_\beta} + \frac{\partial u_\beta}{\partial x_\alpha}) + \frac{1}{2}\frac{\partial \zeta}{\partial x_\alpha}\frac{\partial \zeta}{\partial x_\beta}$. The first term is similar to the strain tensor for small deformations except that it also includes second-order terms for the derivatives of ζ.

Evaluating the two equations that govern the equilibrium of a graphene sheet of thickness t under an external force f per unit area with transverse deformation $\zeta (\gg t)$

$$f = D\Delta^2\zeta - t\frac{\partial}{\partial x_\beta}\left(\sigma_{\alpha\beta}\frac{\partial\zeta}{\partial x_\alpha}\right) \text{ and } \partial\sigma_{\alpha\beta}/\partial x_\beta = 0 \qquad \text{eq. (3.4),}$$

where $\sigma_{\alpha\beta}$ is the stress tensor.

Introducing the stress function χ defined by $\sigma_{xx} = \partial^2\chi/\partial y^2$, $\sigma_{xy} = -\partial^2\chi/\partial x\, \partial y$, $\sigma_{yy} = \partial^2\chi/\partial x^2$, one obtains $\Delta\chi = \sigma_{xx} + \sigma_{yy}$ and the equation for f as

$$f = D\Delta^2\zeta - t\left(\frac{\partial^2\chi}{\partial y^2}\frac{\partial^2\zeta}{\partial x^2} + \frac{\partial^2\chi}{\partial x^2}\frac{\partial^2\zeta}{\partial y^2} - 2\frac{\partial^2\chi}{\partial x\, \partial y}\frac{\partial^2\zeta}{\partial x\, \partial y}\right) \qquad \text{eq. (3.5).}$$

To obtain the equation of equilibrium in terms of the stress function and the deformation we have

$$\Delta^2\chi + Y\left(\left[\frac{\partial^2\zeta}{\partial x^2}\frac{\partial^2\zeta}{\partial y^2}\right] - \left[\frac{\partial^2\zeta}{\partial x\, \partial y}\right]^2\right) = 0 \qquad \text{eq. (3.6).}$$

Equations (5) and (6) form a complete set of equations for large deflections of plates. The set is, however, very complex and cannot be solved exactly. These are non-linear equations and even for simple systems one has to make approximations to get the order of magnitude estimates for the energies and deformations [71]. This is the starting point for the deliberations on such systems and we will evaluate for given deformations the estimates of the forces that may be responsible.

The first estimate is for the situation when dealing with large deformations

$\zeta \gg t$, here, the relation between the deformation and the associated force will be determined. Estimation of the terms in eq. (3.6) shows that $\chi \sim Y \zeta^2$, similarly the first term in eq. (3.5) is smaller than the second which is of the order of magnitude $Yt\zeta^3/R^4$, where R is the radius of the plate. Since this is comparable with the external force, we have

$$f \sim Y t\zeta^3/R^4 \qquad \text{eq. (3.7)}.$$

This is a non-linear relation where force is proportional to the cube of ζ. Comparing equation (3.7) with eq. (3.2), immediately shows that there is a linear relationship between f and ζ for small deformations while it is not the case for large deformations. In addition, we get only approximate solutions in the case of large deformations. For small deformations $f \sim \zeta$ (eq. (3.2)) and for the large ones $f \sim \zeta^3$ from eq. (3.7). These two approximations indicate the order of magnitude of the **bending** versus the **stretching** effects when dealing with the curvature of C nanostructures [70].

3.5. The spherical curvature in Corannulenes

Initiation of curvature in any embryonic C structure due to the formation of a pentagon is the essential starting point for the cage closure resulting in the formation of spherical caps for nanotubes or fullerenes. In figure 3.1(a) and 3.1(b), the top and side views of a corannulene are shown with the two important dimensions; the curvature can be viewed as a spherical protrusion with $r \sim 1 - 2\text{Å}$ in the larger circular curved sheet of $R \sim 3 - 4\text{Å}$ consisting of the hexagons only. For fullerenes with icosahedral symmetry I_h the center of the pentagon has C_5 symmetry [3]. Symmetry arguments can be employed on the formation of either the fullerenes or single-walled nanotubes from the embryonic corannulene-like or coronene-like structures [55]. The figure illustrates the curvature-related stresses and relates the magnitude of protrusion and deformation induced by external forces as $r \approx \zeta$.

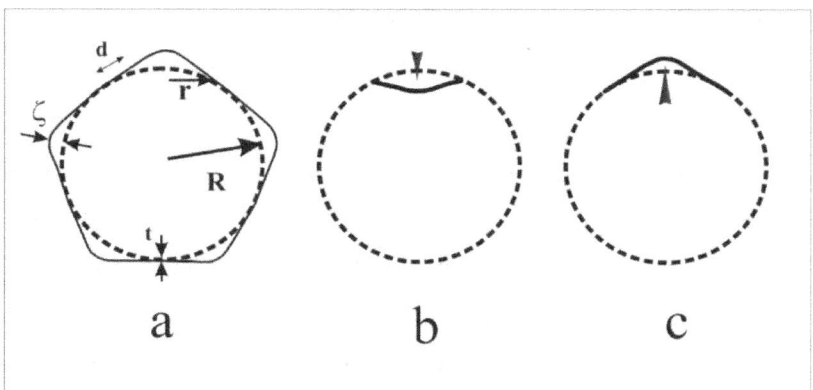

Fig. 3.4. A typical, large (>C_{60}), icosahedral fullerene with the twelve protrusions is shown in (a) with protrusions $\zeta \geq t$, superimposed on a sphere of radius R, thickness t. In (b) and (c) is shown that these protrusions arise by subjecting a perfect spherical shell to a concentrated force f_0 along the inward (b) or the outward normal (c).

A typical fullerene (>C_{60}) is shown in figure 3.4(a) with protrusions $\zeta \geq t$, superimposed on a sphere of radius R, thickness t. These protrusions can be produced when the spherical shell is subjected to a concentrated force per unit area fo along the inward (Fig. 3.4(b)) or outward normal (Fig. 3.4(c)). The resulting deformation $\zeta = R_0 - R$, where R and R_0 are the radii of the inscribing and circumscribing spheres of the respective Goldberg polyhedra. In the case of fullerenes with protruding corannulenes like C_{80}, C_{140}, C_{180} etc, one can determine the size of the pentagonal protrusion ζ from topographical features and relate these to the stresses that are generated. These internal stresses are of the order of the force f_0 of Fig. 4(b) and 4(c). A major part of the elastic energy is stored in the narrow bending strip $\sim d$ on the edge of the bulge. Geometrically, angle α is the angle subtended by the bulge at the spherical center; $\alpha = \zeta/d \approx r/R$; area of this strip $\sim rd$.

The deformation ζ will vary from zero to maximum over the distance d. The inscribing shell's radius R can be considered to be equivalent to the radius of the corresponding fullerene. The distance d is a measure of the region of the protrusion and the area of deformation $\sim d^2$. The pure bending energy over this area varies as $E_{ben} \sim Yt^3(\zeta/d)^2$ as per eq. (3). The **stretching** accompanying the **bending** in flat plates is a second order effect as evident from Eq. (1) but in the case of shells the order is reversed and **stretching** becomes the first order effect. The infinitesimal strain tensor $\sim \zeta/R$ and the corresponding stress tensor $\sim Y\zeta/R$ and the deformation energy per unit area $\sim Yt(\zeta/R)^2$. Over an area d^2 the total stretching energy

$$E_{str} \sim Y t[(\zeta d)/R]^2 \quad \text{eq. (3.8)}.$$

The bending energy decreases and stretching increases with the increase in d thus both energies are considered in determining the deformation near the point of application of the force. Minimizing their sum $E_{ben} + E_{str}$, one gets $d \sim \sqrt{tR}$. As can be visualized from Fig. 3.4, the **bending** is along the meridian and **stretching** along the circle of latitude (of radius r). The total elastic energy in the bending strip of a corannulene is

$$E_{cor} \approx Yt^{5/2}(\zeta^{3/2}/R) \quad \text{eq. (3.9)}.$$

For a given value of the deformation one can obtain the outward force f_0 by equating it to the derivative of E_{cor} with respect to ζ as

$$f_o \approx Y t^{5/2}(\zeta^{1/2}/R) \quad \text{eq. (3.10).}$$

As the bulge or the outward protrusion has occurred due to the generation of internal stresses associated with the curvature of the corannulene, this stress P can be associated with the work done in producing a defect volume $\sim \Delta V$, where $\Delta V \sim r^2 \zeta \sim \zeta^2 R$. The total free energy $= E_{cor} - P\zeta^2 R$. Derivative of this total free energy yields the critical stress

$$P_{cr} \approx Y t^{5/2}/(R^2 \zeta^{1/2}) \quad \text{eq. (3.11).}$$

This is an inverse relation between ζ and P (i.e., ζ increases when P decreases) hence it indicates an unstable equilibrium; therefore, the bulges with large ζ grow on their own accord, while the smaller ones will shrink. Equation (3.11) corresponds to a maximum of the total free energy. The emergence of a bulging curved surface can be related with an internal, outward force f_0 whose magnitude is obtained by taking the derivative of E_{cor} with respect to ζ. The nanoelastic model uses the above mentioned arguments to provide a picture of the typical soot forming environment where C accretion produces planar as well as the curved C structures. It is the curvature related surface forces that play a decisive role in determining whether fullerenes or nanotubes will be formed.

In Fig. 3.5, the upward lifting force per unit area f_o from eq. (3.10), applicable on a corannulene and the associated critical stress P_{cr} (from eq. (3.11)) generated in the shelled structure are plotted as a function of C_x where x is the number of C atoms in the fullerene. The crossover is around C_{240} when the critical pressure for the bulges to grow becomes smaller than the outward bulging force and the structure becomes unstable. Although the figure has been calculated for a complete fullerene with 12 corannulenes, the results are equally valid for a hemispherical cap that will lift and pull an SWCNT underneath.

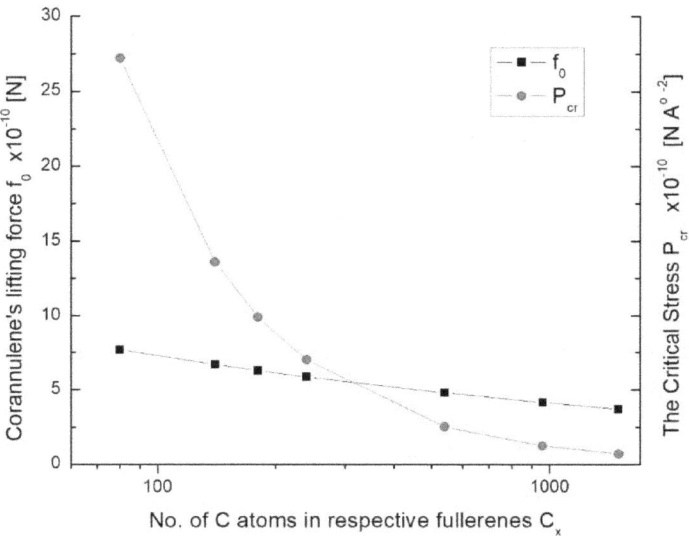

Fig.3.5. The upward lifting force per unit area f_o (equation (3.10)) on a corannulene and the associated critical stress P_{cr} (eq. (3.11)) are plotted as a function of C_x where x is the number of C atoms in respective fullerenes. The crossover is around C_{240} when the critical stress for the bulges to grow < than the outward bulging force and the structure becomes unstable. **Note: large ($\geq C_{60}$) shelled structures are considered here.**

3.6. The Corannulene- and Coronene-based growth models

The fullerene spheroids when treated as Goldberg polyhedra reveal the symmetries of respective fullerenes around the chosen axes of symmetry [3]. The vertices of these polyhedra are given by $n = 20(b^2 + bc + c^2)$, where $b = c = 1, 2, 3, 4, 5...$ for $C_{60}, C_{240}, C_{540}, C_{960}, C_{1500}$. These fullerenes and the higher ones with I_h symmetry, are the only ones that can be shelled inside each other to produce carbon onions with the same mutual spacing which exists within the graphite layers, i.e., 3.34 Å [72,73]. There are other set of icosahedra with similar symmetries, C_{80} and C_{180}, have b = 2,3 and c = 0, that can also grow in the soot. Figure 3.6 shows a set of four icosahedral caps that can grow around a central corannulene shown as the dotted circle at the center. These caps respectively belong to C_{60} (Fig. 3.6(a)), C_{240} (b), C_{80} (c) and C_{180} (d). The characteristic feature of these icosahedra is the way in which the six pentagons of the hemispherical caps are geometrically arranged.

The central pentagon and the circumferential five pentagons are shown in armchair (3.6(a) and (b)) and the zigzag (3.6(c) and (d)) geometries around the same C_5 axis. The C_1, C_2, C_3 addition route is specific for each structural growth. These are shown to depict the pattern of growth out of these structures that may either lead to fullerene formation or the growth of a single-walled nanotube. Once an armchair cap is formed, it grows further as an SWCNT with the addition of C dimers C_2. Each C_2 is then equivalent to a hexagon that is added to the tubule in an armchair configuration. The growth mode would require C_2 addition at the roots of the outward growing nanotube. The requirement for the further growth of C_{80} and C_{180} is the addition of C_1 and C_3, as shown. The growth of the zigzag nanotubes requires a carbonaceous vapour that is equally rich in C_1, C_2 and C_3. Thus an upward growing SWCNT in zigzag mode has a barrier in terms of the statistical requirement of almost equal abundance of the three component species of the growth, i.e., C_1, C_2 and C_3, as opposed to armchair SWCNTs that need only C_2. The four icosahedra presented in Fig. 3.7 are shown around a central hexagon along the C_3 axis.

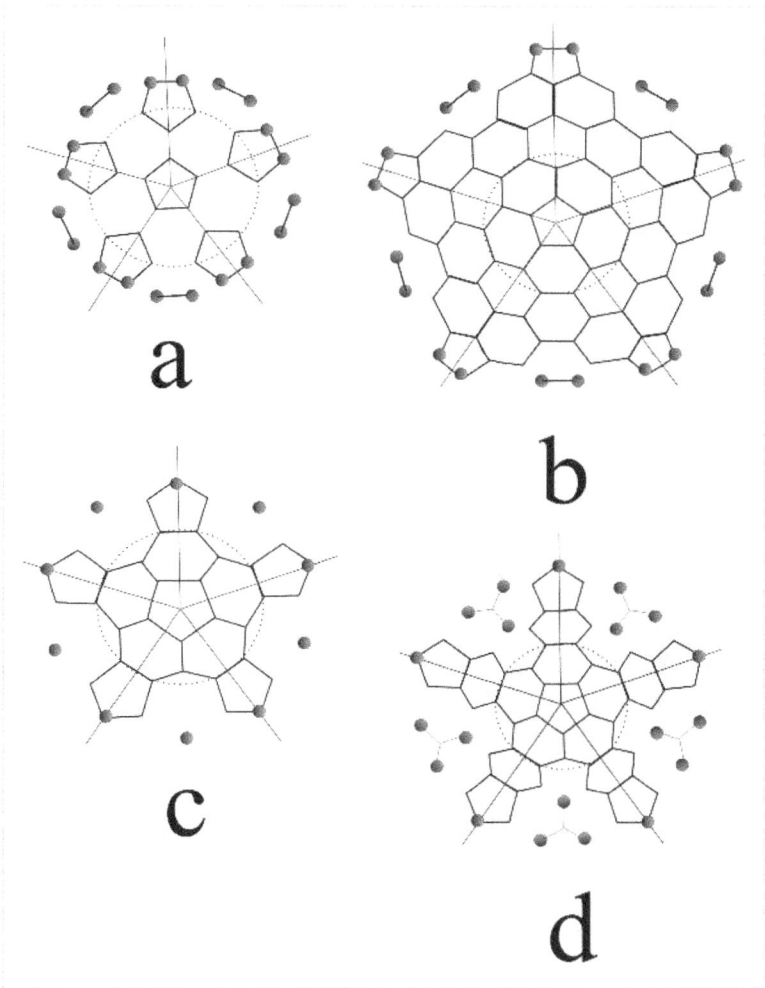

Fig. 3.6. A set of four icosahedral caps are shown that can grow around a central corannulene shown as the dotted circle at the center. These caps respectively belong to C_{60} (a), C_{240} (b), C_{80} (c) and C_{180} (d). The common feature of the four icosahedra is the central pentagon and the circumferential five pentagons in armchair ((a), (b)) and zigzag ((c), (d)) geometries. The C atoms and molecules (C_1, C_2, C_3) are indicated that would be needed for a particular pattern to grow out of these structures that may either lead to fullerene formation or the growth of a single-walled nanotube.

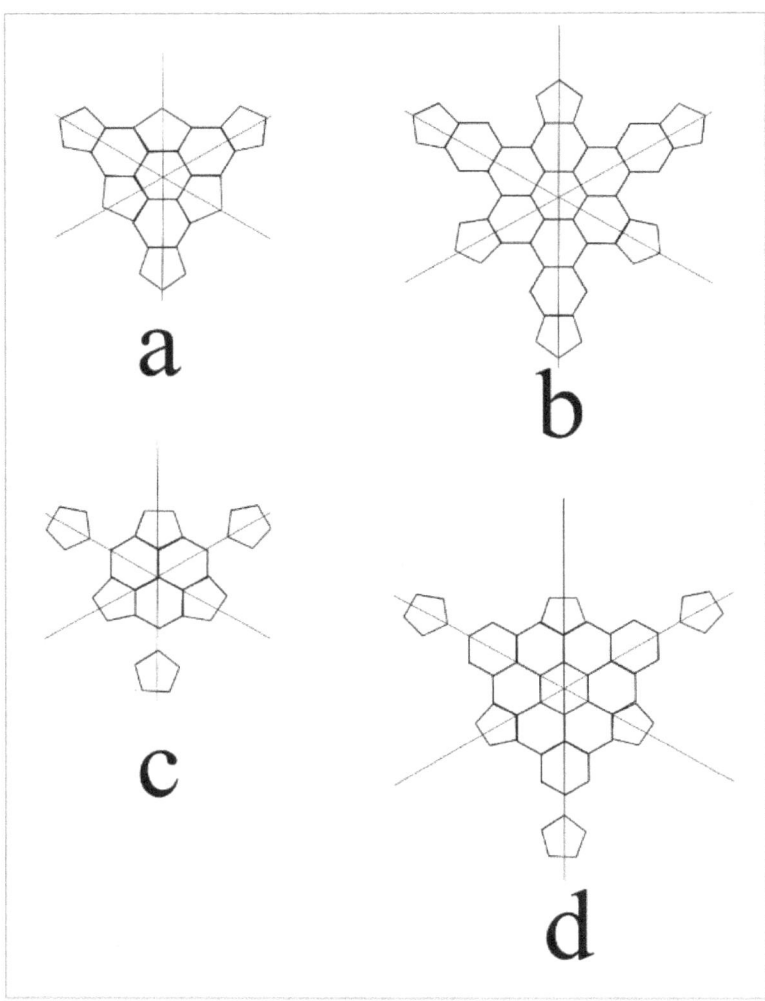

Fig. 3.7. The four icosahedra shown in Fig. 3.6 are shown again around a central hexagon (with the exception of C_{80}) along the C_3 axis. In the case of (a) and (b), a full coronene (a flat hexagon-only bulging block of graphene) does not develop while, in the case of (c) and (d), it develops as the center piece. It can be seen that along C_3 axis the structures belonging to C_{60} and C_{240} produce zigzag geometry while C_{80} and C_{180} produce armchair arrangements.

The common feature of all of the four structures is the dispersal of the two sets of three pentagons each, around the C_3 axis. The peripheral set of three pentagons is in the zigzag configuration for C_{60} (Fig. 3.7(a)) and C_{240}

(3.7(b))), while it produces an armchair arrangement for C_{80} (3.7(c)) and C_{180} (3.7(d)). Therefore, the requirement (of C_1, C_2 and C_3) for the further growth of icosahedra that are growing around the C_3 axis is exactly opposite to that for the two sets shown in Fig. 3.6. In the case of Fig. 3.7(a) and (c), a full coronene (a flat hexagon building block of graphene consisting of seven hexagons) does not develop while, in the case of Fig. 3.7(b) and (d), it may develop as the center piece.

3.7. Corannulene versus Coronene as embryos for the spinning fullerenes or the outward growing SWCNTs

For a corannulene that has been described in the previous section, there exists a strong inward bending moment along the circumference $M_o \approx (1+\nu)D/r$ where D is the flexural rigidity of graphene: $D = Yt^3/12(1-\nu^2) \approx 3$ eV or 5.16×10^{-19} N m; we have used $Y = 10^{12}$ Pa, $\nu = 0.163$ and $t = 1.82$ Å. For the corannulene with $r_1 \sim 2$ Å, this bending moment $M_0 \approx 3 \times 10^{-9}$ N. This introduces a bending stress $6M_0/t^2 \sim 3 \times 10^{-9}$ N Å^{-2}. With the addition of pentagons, the bending stresses set the evolving structure into rotations. The torque on the corannulene of moment of inertia $I_0 \sim 2 \times 10^{-45}$ kg m^2 produces angular acceleration $\sim 10^{25}$ rad s^{-2}. The rotational energy of a 3D rotor is a function of temperature T and the moment of inertia can be written as $E_{rot} = k_B T \times \ln(8\pi^2(I_0 k_B T/2\pi^2\hbar^2)^{3/2})$ [74]. E_{rot} is also $=1/2\ I_0\omega^2$, ω is the angular frequency. A rotating structure at 300 K yields a spinning speed of 10^{12} s^{-1}. These rotating curved structures have higher probability of accreting C_1, C_2 and C_3 from the sooting environment to complete the spheroid. These structures are predicted to be spinning with rotational frequencies $\sim 10^{-12}$ s. The above arguments from the symmetry and the geometric considerations provide the justification for the observations of the spinning C_{60} with rotational frequency~10^{-12} s in the condensed solid form the C_{60}-fullerite [74,75]. There exist a range of values of the parameters Y and t for graphite, fullerenes and the nanotubes used by different researchers [76-79]. Their range of thickness t used or derived from experiments is between 0.7 and 3.34 Å while those of Y are between 5.5 and 1 TPa.

The formation of an initial hexagon can also start the process that can produce shells and nanotubes. Figure 3.7 describes four such situations where the hexagonal net expands until a pentagon is added and the curvature is introduced with the associated bending moment. Symmetry imposes restrictions on the way pentagons can be added to the hexagonal sheet. The spherical curvature is introduced by the two sets of three pentagons each, to complete the hemispherical shell. Both of these sets are displaced by 60° from each other and symmetrically disposed around the center of the initial hexagonal net. In the case of C_{60} and C_{80}, the hexagonal net comprises of one and three hexagons, respectively, whereas a full coronene (seven hexagons) develops in the case of C_{180} and C_{240} before the addition of pentagons. The requirements of symmetry impose constraints on the hexagon-initiated structures to produce the spinning fullerenes as the dominant by-product in sooting environments. SWCNTs are less likely to grow along the C_3 axis due to the rotational torque introduced in two separate stages of three-pentagon addition. Therefore, a closed fullerene shell is more likely to be formed as shown in Fig. 3.7. Right from the moment of the corannulene formation, the earlier addition of pentagons enhances the probability of fullerene formation. Spinning is the essential step that provides a barrier to the formation of nanotubes and it is overcome when addition of the remaining pentagons is delayed by the formation of a hexagonal network around the central corannulene as in

Figs. 3.6(b) and (d). In the case of the C_{240} cap shown in 3.6(b), thirty hexagons surround the central pentagon in the form of three rings. The addition of the remaining five pentagons takes place at the periphery. The bending stress introduced and uniformly distributed by the circumferential pentagons is about 7 Å away from the central pentagon and may introduce an upward lift to the cap as opposed to the spin in the case of the smaller ones. The case for the C_{180} cap shown in figure 3.6(d) is equally interesting and worth considering as a candidate for the preferential growth of a zigzag SWCNT with 12 Å diameter and not acting as a seed for the growth of a fullerene for the same reasons. Here again the addition of 30 hexagons around the central pentagon provides the delay in the pentagon-induced bending stresses until the addition at the circumference occurs. In this case the resulting cap is in zigzag geometry and the only difficulty for the growth of this cap is the requirement of addition of C_1 and C_3 to provide for the necessary tubule formation, as opposed to the requirements of the addition of C_2s in the growth of armchair SWCNTs.

A note of clarification is in order: When addition of a pentagon or hexagon was referred to in the earlier discussions, it was implied that accretion of C_1, C_2 and C_3 led to the pentagons and hexagons being incorporated into the growing structures. The self-organization among the nanostructures is rooted in the soot that occurs by the accretion of C atoms as the vertices of the fullerenes and nanotubes in a wide variety of isomers, shapes and configurations.

3.8. The ½ fullerene caps of SWCNT

The sole surviving, stable isomer of C_{60} has the icosahedral symmetry I_h. There are a total of 1812 equivalent closed cages of 60 vertices, 12 pentagons and 20 hexagons, but only one has the icosahedral symmetry I_h. The structural stability of the surviving isomer depends upon the equal distribution of the steric strain of the closed cages is ensured by the non-abutting pentagonal configuration [3,80]. The outward lift introduced by the pentagonal protrusion start with an initial curvature $r_1 \sim$ 2 Å; the addition of hexagons stabilizes it by distributing the stress in a larger area, ~ 51 Å^2.

The fullerenes smaller than C_{60} have adjacent pentagons and have larger curvature-related stresses. The associated strain is not uniformly distributed in the polyhedral structures that do not display icosahedral symmetry. All of the 12 pentagons are abutting in C_{20} and the resulting strain is maximized in the shell of diameter 4 Å. C_{20}, therefore, is a highly strained structure and an example of an inherently unstable fullerene due to the pentagon-induced strains, whereas in the case of C_{60}, the 12 non-adjacent pentagons form a perfect spherical shell that possesses the unique distinction of sharing all 60 C atoms uniformly in its 12 pentagons, and 20 hexagons with the 12 corannulenes emerge as the byproduct of sharing all sets of five atoms producing pentagons in such a way that these are used up to form the surrounding ring of five hexagons. Non-abutting pentagon rule is equivalent to the abutting corannulenes I_h C_{60} and the curvature-related strain is uniformly distributed over the spherical shell.

Shelled structures with radii larger than 3.5 Å have interpenetrating corannulenes with increasing separation of the pentagons and the resulting bending stresses are distributed in larger areas. The description of the structure growing around a central corannulene in the previous sections provides a measure for the two regions of the concentration of the **stretching** and **bending** energies for the outward protruding pentagonal defects. The outward lift in a corannulene results from the **stretching** of a region of curvature r_1 and area ~14 Å². The next stage is the addition of the C atoms from the surrounding C vapour that introduce either pentagons or hexagons around the central, lifting pentagon. The addition of hexagons stabilizes it by distributing the stress in a larger area ~51 Å².

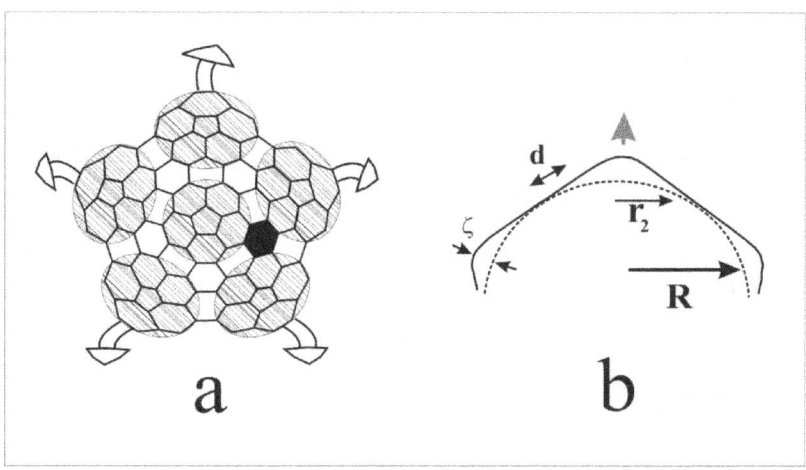

Figure 3.8. (a) The hemispherical cap belonging to C_{240} of diameter 14 Å is shown. The six non-abutting corannulenes are shown as hatched areas; the five arrows along the circumference indicate the bending moments as well as the direction of growth of the structure dependent upon the addition of dimers (C_2). The net outcome is the armchair SWCNT. A blackened hexagon indicates the option of growth around a coronene as shown in (b). (b) The same structure as in (a) but shown along the direction of upward growth indicated by the heavy arrow. The four typical dimensions are the same as in Fig. 3.1.

Fig. 3.8(a) shows the armchair cap belonging to C_{240} of diameter 14 Å is shown with the six corannulenes as the hatched areas. Each corannulene is surrounded by a ring of 10 hexagons. This is the **bending** strip shown as the region d in Fig. 3.8(b) in which the bending energy is concentrated. A blackened hexagon in (a) indicates the alternative option of growth centered on one of the five possible coronene. Such geometry is shown in

Fig. 3.8(b) where a zigzag cap is formed. Here, the same structure as in 3.8(a) is shown along the direction of upward growth indicated by the heavy arrow. The four typical dimensions are similar as in Fig. 3.7(c). However, the 14 Å hemispherical half caps require 90 C atoms to form the 6 pentagons and 30 hexagons as shown in Fig. 3.8(a). To complete the half-cap of C_{240}, 120 C atoms are needed that would yield the six non-abutting corannulenes. Fig. 3.8(a) shows such an armchair half-cap belonging to C_{240}. These non-abutting corannulenes are shown as hatched areas; the five arrows along the circumference indicate the bending moments as well as the direction of growth of the nascent structure. The net outcome may be an armchair SWCNT or a fullerene, depending upon (i) the addition of C_2s only, or (ii) a mixture of C_1, C_2 and C_3, respectively. Cage closure to produce an icosahedral C_{240} fullerene, however, is less likely when the structure is rising from a sooted surface with the probability of 1 in $\sim 10^5$ isomers!

The next most stable fullerene, growing out of the condensing C vapour is C_{70}. The emergence of C_{70}, as a smaller fraction of the C_{60}, provides experimental evidence in favor of the above analysis. It was elaborated earlier that the absence of the non-abutting corannulenes may lead to the stopping of the growth of this (the C_{70}) as smallest and the second-most prominent member of the fullerene family into a nanotube of diameter 7 Å and length > 9.5 Å. C_{70} comprises of the two halves of C_{60} separated by a ring of five hexagons added in an armchair configuration. For the same reasons, the C_{70} can also be considered the smallest, most stable nanotube-like structure that grows in environments that favor the formation of fullerenes [3]. The broad definition of fullerenes shows C_{70}, as a fullerene in symmetry group D_{5h}, as a symmetric top that has evolved out of an icosahedral cap belonging to the spherical top category. The curvature induced by the six abutting corannulenes in the half-cap of C_{60} is stabilized by the addition of five hexagons that share their strain with a similar half-cap to produce a highly symmetric single-walled nanotube where the elastic strains have refused to allow the further growth of the nanotube. Our criterion for the spinning structures provides justification for the dominance of fullerene formation as opposed to that of the nanotubes within the C vapour.

3.9. The summary of Chapter 3

The nanoelastic model developed and applied to the spherical closed cages has elaborated:

(1) The importance of the role of the spherical curvature induced by pentagons in the embryonic structures.

(2) The identification of the conditions that lead to the formation of either a fullerene or a nanotube.

(3) The explanation for the dominance of C_{60} among the fullerenes.

(4) The reason for the observations of the dominance of 14 Å diameter armchair nanotubes, i.e. a SWCNT with a half-C_{240} as the cap in the carbonaceous environments of the self-organizing soot.

The Self-Organizing Soot

Chapter 4

The C_2 Gas

4.1 The C_2 gas

The dynamic role of the C_2 gas was proposed and discussed in various types of soot forming environments [80-84]. Chapters 1 and 2 presented and discussed the experimental evidence of the nature and the diversity of C clusters in the carbonaceous discharges and the dynamic constitution of the regenerative soot. The formation and fragmentation of the linear chains, cyclic rings, graphene sheets and the closed cages as a function of the parameters of the regenerative soot were discussed. The presence of and, the relative contributions of C_1, C_2 and C_3, in the emerging soot were considered in detail in Chapter 2. These were shown to be the basic constituents of the regenerative soot. The experimental data obtained by utilizing different techniques identified the two essential sequences that determine and define the closed cage forming environments. It implied that the transition from the C_3-dominated discharges to the closed cage environments in a C_2-dominated regenerative soot is the necessary condition for the formation of fullerenes. In addition, the C_2 gas emerges as the essential component and the by-product of the dynamic regenerative soot where the closed cages form and fragment. The present chapter highlights the role and contribution of the C_2 gas in the emergence and sustenance of the regenerative soot and as an agent of the self-organization of the cages.

Molecular dynamics simulations have also been employed to investigate C_2's role in the formation of the closed cages [85-90]. Results from an MNDO simulation are shown in Fig. 4.1. The simulation was

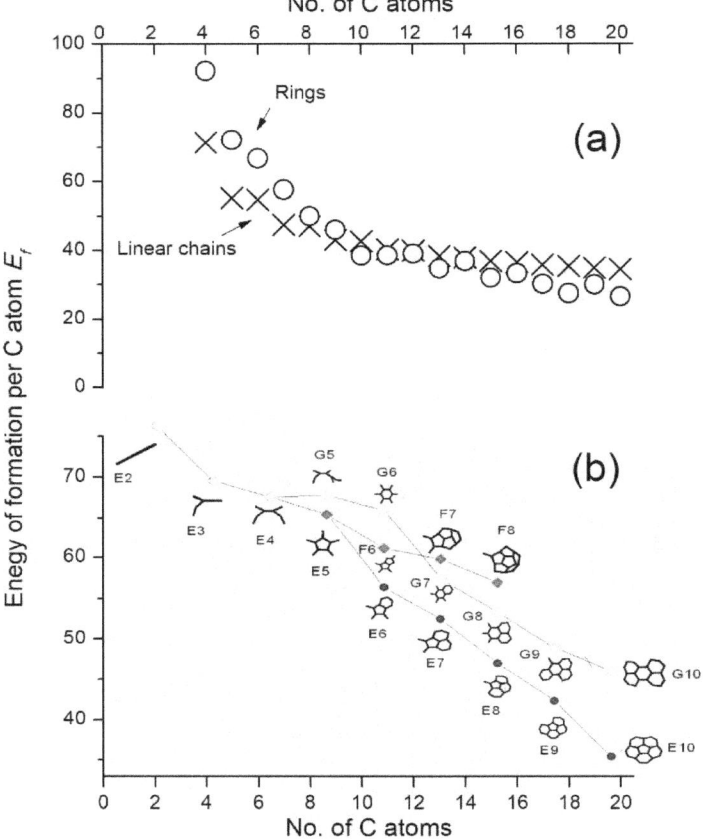

Fig. 4.1. Energy of formation E_f as function of number of C atoms. (a) Addition of C_1s leading to linear chains and ring. Odd and even numbered structures are formed. (b) Simulation results with sequential addition of C_2s in the regenerative soot. The C_2-addition stages are shown for the graphene sheets from G5 to G10; pentagon-based corannulene from E5 to E10 and the all-pentagons steps F6, F7 and F8. Corannulene emerges as the structure with the lowest formation energy and the highest probability. Data from ref. [90].

done to evaluate the energies of formation of clusters of all shapes and forms in the regenerative soot as a function of the increasing number of C atoms per cluster. Chapter 1 showed that the irradiated graphite emits clusters in the variable relative ratios that change as a function of the irradiation parameters. Figures 1.1 and 1.2 showed the evidence and discussed the emergence of the soot with variable relative densities of its basic constituents C_1, C_2 and C_3. These constituents become the feed material of the soot which has a regenerative character that was illustrated in Chapter 2. The mass and the emission spectroscopies of the soot provided the experimental evidence for the existence of the linear chains, cyclic rings, sheets, the open and closed cages.

The energy of formation E_f of the linear chains and rings, per C atom, are shown in Fig. 4.1(a). The figure shows that the cyclic rings are energetically favorable after C_x; $x \geq 10$. The figure illustrates the pattern of clusters with <20 C atoms. However, in the case of clusters C_x; $x \geq 20$, as it was discussed in Chapters 1 and 2, that the mass spectra of the species emitted from the well-sooted discharges is dominated by the closed cages for $x \gtrsim 36$ C atoms. Additionally, the data of the emission spectroscopic measurements of C_2's molecular emission Swan band suggested that C_2 is as the most active constituent of the regenerative soot. Therefore, the simulation of the formation of clusters with the addition of C_2s was also carried out and the results are shown in Fig. 4.1(b). The results are shown for the clusters in the range from 2 to 20 C atoms, for comparison with results of C_1 addition-route in Fig. 4.1(a). Similar trends were seen and confirmed by other researchers [85-89].

The scale of E_f in the two comparative figures (a) and (b) and the pattern of the emerging clusters suggest that the pentagon-assisted, curved structural growth becomes energetically favourable than the all-hexagon sheets for the increasingly larger C clusters.

4.2. The fullerene isomers I_x

Fullerenes as a class of the closed cages of the twelve omnipresent pentagons and a varying number of hexagons have more than one way of arranging these on the spheroid's surface leading to multiple isomers per fullerene. Atlas of Fullerenes [3] provides a compilation of the possible isomers for each fullerene. The isomeric distribution of fullerenes as a function of x indicates an increasing exponential dependence as shown in figure 4.2. The exceptions include C_{20} that has no hexagon and hence the twelve abutting pentagons assemble as a dodecahedron, C_{24} and C_{26} each with 1 isomer.

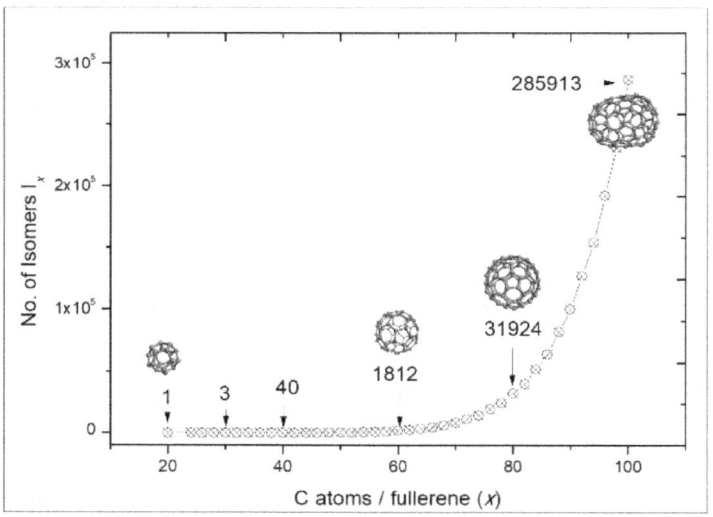

Fig. 4.2. Fullerene isomers I_x increase exponentially with $x \propto exp(l_0 x)$, ; l_0 is an exponent associated with the dynamics of the multiplicity of fullerene structures with the same number of C atoms i.e., isomers.

Onwards from C_{28} with 2 isomers, the multiplicity of the isomers for C_{40} induces exactly 40 isomers, the number increases to 271 isomers for C_{50} and to 1812 for C_{60}. Larger fullerenes have exponentially increasing number of isomers I_x for each C_x. Fig. 4.2 has the range from $x=20$ to 100 to illustrate the increasing trend of I_x versus x that explains the dominant presence of the closed cages with $x > 20$ in the regeneration processes in the soot. An exponential distribution is fitted to the isomeric data that yields $I_x \sim exp(l_0 x)$, where l_o along with x, characterizes the distribution of the respective isomers. The increasing isomer density for larger values of x is equivalent to the probability of the existence of the equally abundant numbers of the respective fullerenes.

4.3. Dynamics of the bottom-up cage formation

Figure 4.2 demonstrates that carbon vapour condenses with higher probabilities for clusters C_x with the increasing number of carbon atoms-x. Linear chains, rings and sheets have single isomer for each type until $x \geq 20$ when the closed cages start to form and thereafter, have the probability of the increasing number of isomers for each x. The cluster growth dynamics of the initial, bottom-up phase is dominated by the closed cages whose respective heats of formation E_x reduce with the increasing x. The topological requirement for the cage's formation energy to reduce for the larger ones ensures higher densities of the closed cages with the increasing spheroidal volumes. The decreasing energy of formation of the cages lead to the probability of their formation being proportional to $exp(-E_x/kT)$. The cluster forming carbon vapour environment modeled by MNDO technique with the C_2 addition route to the formation of cages is shown in Fig.4.3 which extends the data presented in Fig. 4.1 to the possible mechanisms for the formation of the cages. The formation energy per carbon atom E_x is plotted for all the clusters, open and closed cages between $x = 2$ and 60. The formation energies of the open cage route versus the closed cage road are also shown in the figure. The increasingly larger open and closed cage clusters have subsequently, lower formation energies, and the trend continues for the higher fullerenes. Again it can be seen that the trend of the increasing isomers per cage $I_x \propto \exp(l_0 x)$ continues well beyond the C_{60}. The higher probabilities for larger fullerenes to form in a condensing carbon vapor dominate the emerging grand canonical ensembles with the higher proportion of the larger fullerenes. The mechanisms for the transformations into the smaller and eventually the perfect spheroidal neighbor-Buckyball, are explored in the later sections.

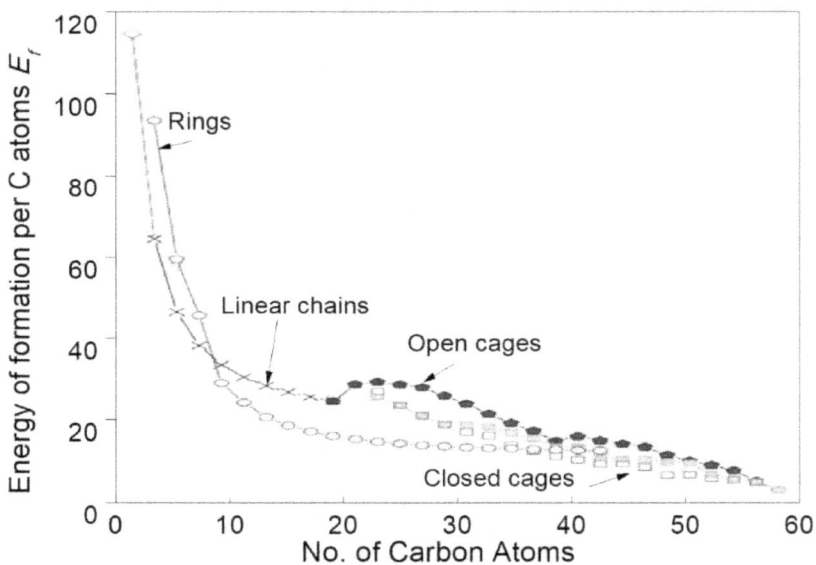

Fig. 4.3. Results from the MNDO simulations for the C-addition in the regenerative soot are shown for all clusters from 2 to 60 C atoms. Energy of formation per C atom E_x is plotted as a function of the number of C atoms x, for $x=2$ to 60 are shown, The cage closure and the stable isomer production has the lowest E_f. (adapted from ref. [90])

The twin conditions, the first of the reducing E_x as shown in Figures 4.1 and 4.3, and secondly of the increasing number of isomers per fullerene-the closed cages in Fig. 4.2, imply that the probability of formation of closed cage fullerenes with x C atoms

$P(C_x) \propto \exp(-\beta E_x + l_0 x)$ where $\beta = 1/kT$.

Therefore, the larger the cluster, the higher the probability that it will be a closed cage with the increasing number of isomers. The net effect of the factor $\exp(-\beta E_x + l_0 x)$ on the condensing C vapor is the higher proportion of the large fullerenes. The other essential ingredient is the diatomic molecule C_2 which emerges as the agent of transformation; when it is added to a cluster, the cage grows to be the next higher one

$$C_x + C_2 \rightarrow C_{x+2};$$

if C_2 is excluded then a one-down cage is formed

$$C_x \rightarrow C_{x-2} + C_2.$$

The time-of-flight spectra from ablated graphite have the higher fullerenes peaked around ~100 to150 C atoms, with the lower end around ~40 C and the maximum determined by the detection capabilities of the mass analyzer [2]. Similarly, the Chapters 1 and 2 provided the evidence of the larger cages' dominance of all of the mass spectra extracted from the regenerative source. Clearly, an entire range of cages is formed followed by the cycle of growth and shrinkage under the constraints of the pentagon-induced elastic strains. Therefore, the higher probability for the large C_x ($x > 60$) cages is inherent to the cage formation mechanisms, and in addition, the formation of the large, unstable cages is the basic premise for the self-organizing processes in the regenerative soot. The fragmentations initiate the consequent top-down sequences of the shrinking cages.

4.4. Dynamics of the top-down cage shrinkage

The foregoing dynamical description of the growing fullerenes of the ever increasing sizes and higher isomer densities introduces the two essential stages; the first is the exponentially growing cage population of the ever increasing C-content and diameters and the second is the perceived process of the growth of cages via C_2 addition which requires a proficient source of C_2s. The accumulative C_2 gas for the formative process emerges out of the fragmentations. Its presence was explained by the experimentally observed compositions of the regenerative soot in Chapter 2. The increasing number densities of the larger cages due to the availability of their respective isomers I_x, defines the sooting environment where cage↔cage collisions, cage→wall and the cage↔C_2 interactions and the nonlinear surface forces associated with the local and global curvature of the large fullerenes generate a top-down series of the shrinking of the cages. The bottom-up growth of fullerenes with ever increasing C-content, diameters and the pentagon-related local deformations is accompanied by the top-down cage fragmentation sequences. The surface deformation-induced strain, discussed in Chapter 3, is partially released with the emission of a C_2 at each fragmentation step $f: C_x \to C_{x-2} + C_2$.

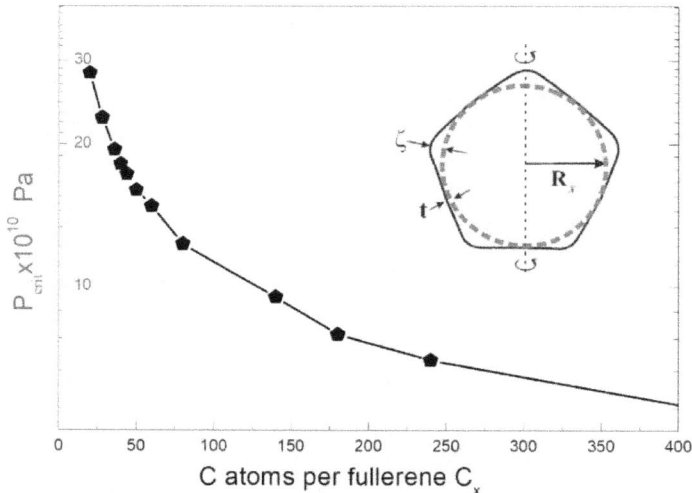

Fig. 4.4. The cage stability criteria are shown as a function of x using the continuum elastic model [55]. The critical stress $P_{crit} \propto \zeta^{-1/2} + R_x^{-2}$, has a nonlinear dependence on ζ. **Inset:** A closed cage with radius R_x, thickness of σ- shell t, and ζ -the protrusion. The fullerene is assumed to be a 3D rotor.

The nano-elastic model developed for the σ-bonded, nano-cages of C, demonstrated the dominance of the stretching effect over the corresponding bending effects, on the elastic response of the C nano shells that generate the pentagonal protrusions [55]. It showed that the deformations in nano spheroidal shells create internal stresses that have a direct bearing on their structural stability. Curvature-related properties were obtained for the stability of a range of fullerenes from C_{60} to C_{1500}. The critical stress P_{crit} was evaluated beyond which the structure becomes unstable. The critical stress has nonlinear dependence on the deformation parameter. This crucial feature of large fullerenes is the direct result of the pentagon-induced, spherical curvature. Fullerenes-the graphene shells of constant thickness t but with variable radii R_x have nonlinear forces associated with curvature ($1/R_x$) of the surface protrusions. The nano-elastic model of fullerenes with pentagonal protrusions superimposed on a spheroidal structure presents two types of curvatures; the first dealing with the local, pentagon region and the second with the global fullerene spheroid. Curvature being inversely proportional to the radius has larger strain- induced surface forces around the pentagons shown in Fig. 4.4. These local stresses generate instability. The critical stress is estimated as $P_{crit} \propto \zeta^{-1/2} + R_x^{-2}$, where ζ is a measure of the local protrusion, R_x–the radius of the respective fullerene –the global curvature of the cage. P_{crit} has a nonlinear relationship with ζ yielding an unstable equilibrium.

This local pentagon-induced curvature provides the inbuilt structural forces that limit the cage size by initiating the shrinking process. The earlier process of the ever increasing, C_2-accreting cages is checked and halted by the surface deformation related forces. The bottom-up grown structure may enter into the regime of instability and the top-down trimming start for the over-grown, pentagonal protrusion ridden closed cages by the emission of C_2 via the cage shrinkage route $C_x \to C_{x-2} + C_2$. The essential additive and reductive component is the C_2. Amongst the two well-known fullerenes with the similar, icosahedral symmetry C_{20} and C_{60}; only one isomer with icosahedral symmetry I_h exists in the case of C_{20} i.e., 1 out of 1 isomer; for C_{60}, the only 1 out of 1812 isomers. The two fullerenes have very different surface topographies with only the icosahedral C_{60} having uniform distribution of the strain due to curvature of the perfect spherical shell.

Non-abutting pentagons are the main reason for the uniform distribution of strain that is not the case for C_{20}. Surface deformations of all other isomers of C_{60} lead to non-uniformity of the structural strain. High symmetry is one of the fundamental conditions with greater probabilities of the spheroidal cages' survival in hot carbon vapour.

The density of C_2 increases as a function of f until

(1) each cluster has either been transformed into a C_{60} and $(x/2 - 30)$ C_2 molecules through a successive, cumulative fragmentation sequences $C_x \to C_{60} + (x/2 - 30)C_2$.

(2) The other option of the top-down sequence may lead to the total destruction of the cage through successive fragmentations $C_x \to C_{x-2} + C_2$.

Both of the options lead to the enrichment of the C_2 gas. Dynamics of the top-down fragmentations can be mapped by the grand canonical ensemble of large cages and the associated C_2 gas.

4.5. The shrinking cages in the pulsed soot

The regenerative soot can be considered as a dissipative self-organizing dynamical system. The initial sequences of the introduction of C clusters by laser ablation, arc discharge or by sputtering of hollow cathodes of graphite, generate the dynamical system which operates by the addition and removal of atoms and molecules. The emerging pulse of the laser ablated plumes of C clusters exist and operate in a finite time scale of ~ns to μs. Arc discharge and regenerative soot are its continuous analogue. A model is developed by using the pulsed soot case and will be extended to the continuous in the next section.

The energy dissipative processes of the formation of large cages followed by their fragmentation generate entropy which characterizes the nature of the net outcomes. A statistical mechanical model describes such a dynamical system [80]. The model works out the details of the self-organizing, grand canonical ensemble in terms of its two components; the first is the calculation of number densities of the forming and fragmenting cages and, the second is the evaluation of the increasing number densities of the C_2 gas that emerges as a consequence. The model is based on three assumptions that:

(1) the laser ablated graphite produces a pulse of closed cages in the same ratio as that of their respective isomers i.e. $N_x \sim I_x$.

(2) The range of fullerenes considered is between C_{60} and C_{100}.

(3) At each fragmentation step half of all the cages fragment through $C_x \rightarrow C_{x-2} + C_2$. C_{60} being the only exception and it acts as an attractor.

Figure 4.5 shows results from 45 successive fragmentation steps of 3D evolution of the distribution of the isomer-based cage population $N_x(f)$ for the respective fullerenes. The original cage distribution in the primary ensemble is at $f = 0$, the number of each of the fullerenes C_x is according to its number of isomers I_0. With each successive f, each fullerene's population reduces at the fixed rate r=1/2. The larger ones fragmenting to enrich the next smaller neighbor's population following

$$f: C_x \rightarrow C_{x-2} + C_2.$$

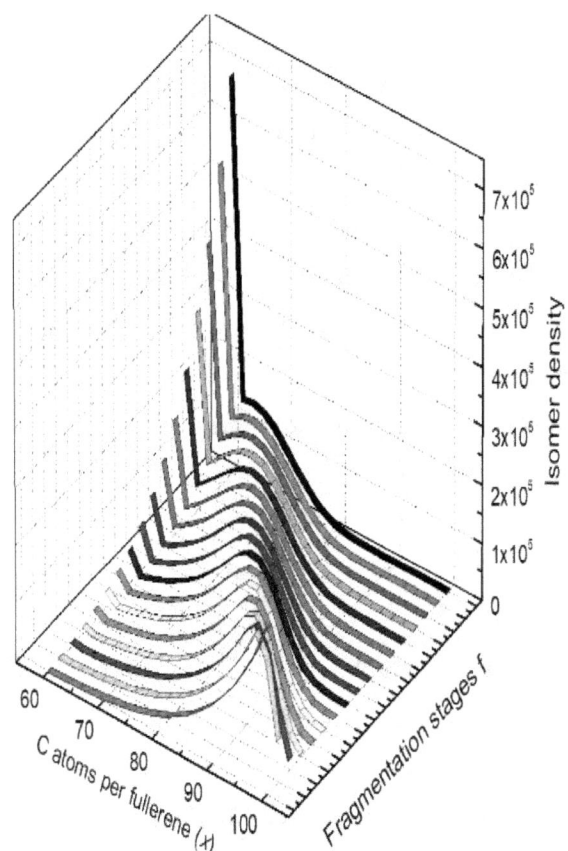

Fig. 4.5. The Initial ensemble of fullerenes from C_{100} to C_{60} at $f=0$ is exposed to the cage shrinking environment. The 3D plot of the dynamics of cage shrinkage via $C_x \to C_{x-2} + C_2$ shows self-organization of the fullerenes, except C_{60}, with half of all cages fragmenting at each step. The starting cage distribution in the primary grand canonical ensemble is $f = 0$. At each successive f all larger fullerenes shrink by spitting out a C_2 per step. Fig. from ref. [81].

The process continues following the Logistic equation $N_{f+1} = rN_f$, where r is the rate of fragmentation per step. We use $r = 1/2$ for the simulations.

At the start of the fragmentation sequence, $N_{f=0}(C_x) = I_x(C_x) = I_0$ where I_0 the total number of each fullerenes C_x at $f = 0$.

At $f = 1$; $(I_0/2)C_x = (I_0/2)C_{x-2} + (I_0/2)C_2$,

At $f = 2$; $(I_0/4)C_{x-2} = (I_0/2)C_{x-4} + (I_0/4)C_2$...

And so on, with all the larger fullerenes shrinking by spitting out a C_2 per step per cage.

The cages shrink as $C_x \to C_{x-2} \to C_{x-4}$ ... with the larger ones reducing ½ of their population at each fragmentation step and the smaller ones gaining sequentially. The C_2 population increases at each step.

By $f = 41$, almost all larger cages can be seen shrinking towards the attractor i.e. C_{60}. For each larger cage ($> C_{60}$) only one cage and $(x/2 - 30)$ C_2s are emitted. The density N_{C_2} of the accumulating gas of C_2 as a function of f shows monotonous increase. However, the total fullerene density is preserved in the model. The rate of increase of $N_{C_2}(f)$ reduces to zero for $f > 40$ when majority of the cages have been transformed into C_{60}. In this conservative model, the total fullerene number density remains constant while the larger cages ($> C_{60}$) transform towards C_{60} and the C_2 gas.

Fig. 4.6 shows the set of fullerenes that were chosen for the simulation of the dynamics of cluster fragmentation of the primary fullerene ensembles formed in a pulsed ablation experiment to illustrate various sequences and stages of the shrinking and C_2-spitting cages had the starting number densities of the respective fullerenes e.g., C_{100}, C_{80} and C_{70} according to their respective isomers 285913, 31924 and 8149. In Fig. 4.6(a), C_{100} population can be seen to reduce significantly within the first 5 steps while that of C_{80} goes through a peak around the 18th step and then diminish by the 30th. The larger fullerenes gradually disappear by enriching their next smaller counterparts. The C_{62} density continues to rise up to $f \sim 35$ and then reduces and acts as a one-way "gate" for the increasing density of C_{60} which is the sole survivor and attractor of the dynamical system. After about 50 steps the final C_{60} population ≈ the sum of all the isomers of the primary fullerene ensemble $\sum I_x$. In our assumption of $60 \leq x \leq 100$ yielding $\sum I_x \sim 3.26 \times 10^5$, therefore, that is the maximum possible number of C_{60} per ablation. The density of the C_2 gas continues to rise to ~ two orders of magnitude higher densities than the cumulative $\sum I_x$ as shown in Fig. 4.6(b).

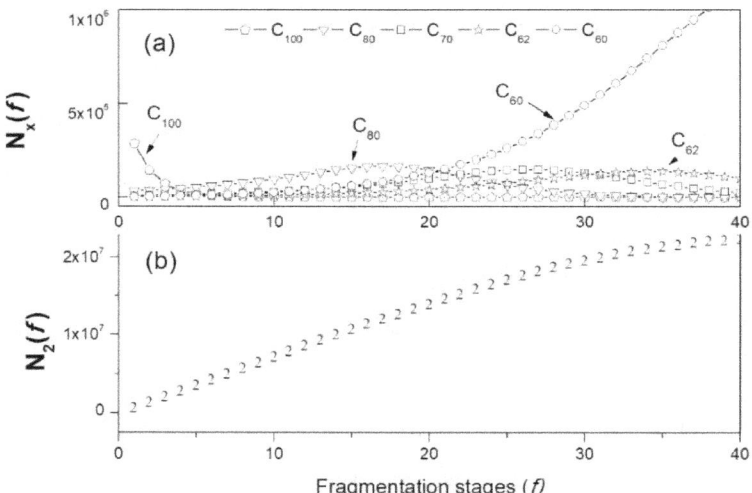

Fig.4.6. (a) Number densities of C_{100}, C_{80}, C_{70}, C_{62} go through a maximum and then reduce to zero within 45 fragmentation steps. The exponential increase of C_{60} density as a function of f is $N_{60}(f) \sim \exp(l_1 f)$; l_1 being a stability index for this dynamic sequence. (b) The cumulative number density of the C_2 gas as function of the fragmentation stage f. Each fragmentation step is equivalent to a C_2.

The simulation assumed that the fragmentation, the shrinkage and conversion processes of all fullerene cages > C_{60} stops at C_{60}. In real life experimental conditions, that may not be the case, the process of fragmentation might continue for some of the cages < C_{60}. The model and its results in Fig. 4.6 indicate the mechanisms that may be responsible for the emergence of C_{60} in a self-organizing regenerative soot.

The pattern of change of population densities $N_x(f)$ of the five chosen fullerenes out of the grand ensemble, the C_{100}, C_{80}, C_{70}, C_{62} and C_{60} shows that all species, except C_{60}, go through a maximum and then reduce to zero within 45 fragmentation steps in Fig. 4.6(a). An exponentially increasing function $\exp(l_1 f)$ can be fitted to the profile of $N_{60}(f)$ plotted as a function of f; l_1 being the coefficient of stability for this dynamic process for the exponential increase of the density of C_{60}.

In this section, we treated the fragmentation via the Logistic relation $N_{f+1} = rN_f$ with $r = 1/2$, for $f: C_x \rightarrow C_{x-2} + C_2$. This was done for the dynamics of the top-down shrinking cages.

4.6. Entropy of the fragmenting fullerenes as rotors

As discussed in Chapter 3, the addition of the pentagons induces curvature in flat graphene-like sheets during the initial growth phase. Alternatively, if a pentagon is the starting structure, then a corannulene bowl with the inherent curvature results; its further growth may yield a cage. Bending moment associated with the curved, bowl-like structure will set it into rotation. Carbon accretion thereafter, takes place into the rotating, curved structure until it closes. Before, during and after the cage closure, all components of the regenerative soot are the rotating structures with their respective rotational moments of inertia. These rotors have rotational moment of inertia $M_x = 2mR_x^2/5$, where m is the mass and R_x the cage radius. In the ensemble of rotating fullerenes, the rotational partition function for each of the fullerene with mass m and rotational moment of inertia with M_x is $z_{rot} = \left((8\pi^2 kTM_x)/\hbar^2\right)^{3/2} = T/\theta_{rot}$, with the rotational temperature $\theta_{rot} = \hbar^2/2M_x k$. The associated entropy of the rotating fullerene is $S_x = kln([(z_{rot})^x]/x!)$ [91]. The rotation of the primary ensemble's fullerenes allows the calculation of their entropies during the subsequent fragmentation sequences. Sum of entropies of all the fullerenes as a function of the fragmentation steps $\sum S_x(f)$ reveals the dynamical features of the self-organizing cages. For all fullerenes larger than C_{60}, two mechanisms compete to reduce the structural strain; one occurs via SW transformation [92] for the re-arrangement of C bonds while the other is the route of fragmentation discussed in the previous section $C_x \rightarrow C_{x-2} + C_2$ with the emission of a C_2. The rearrangement of bonds redistributes spheroid's strain and the processes are energy consuming for all isomers. Intuitively, a typical cage will go through a sequence of steps including both of these options.

Fragmentation not only reduces the surface area but also leaves the cage with fewer atoms to be rearranged. The process is presumed to continue in the primary ensembles of the hot, collision dominated, internal stress-ridden fullerenes. The fragmenting cages are imbedded in a hot gas of C_2s whose density increases with each fragmentation stage. With C_2 linked with every fragmentation step, its increasing density $\propto$ the decrease of the cages' surface area. The hot C_2 gas of growing density is an integral component of the regenerative, grand canonical ensemble. The statistical mechanical treatment of the cages and the C_2 gas as a function of the fragmentation stages describes the self-organization whose outcome is the deformation-free, perfect spheroidal C_{60}. Mapping of the dynamics of the grand canonical ensemble as a function of the cage fragmentation reveals that the positive entropy of the ensuing C_2 gas compensates for the shrinking cages' negative entropy. The total, positive entropy of the fragmenting fullerenes and the C_2 gas during fragmentation stages continuously increases. The cumulative entropy of various sets of fullerenes, however, decreases.

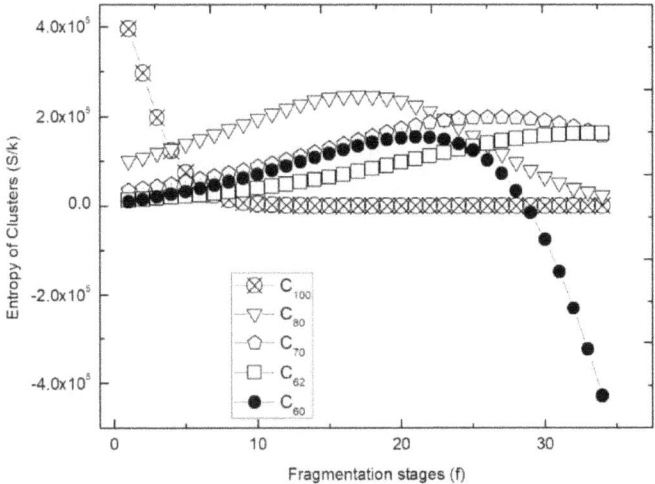

Fig.4.7. The sum of entropies of each of the fullerenes $\sum S_x(f)$ from C_{100} to C_{60} show a net decrease with the increasing fragmentation steps f. The graphs shows variation of $\sum S_x(f)$ for the five selected fullerenes C_{100}, C_{80}, C_{70}, C_{62} and C_{60}. $\sum S_{60}(f)$ becomes negative for $f \geq 28$ and stays negative. **Note:** The numbers of fullerenes follow the fragmentation profiles shown in Fig. 4.6(a).

In Fig. 4.7, the variations of the rotational entropies $\sum S_x(f)$ of all isomers of the five selected fullerenes provide insight into the details of the fragmentation dynamics with $\sum S_{100}(f)$ reducing to zero within the first 10 fragmentation steps while that of C_{80}, C_{70}, C_{62} and C_{60} first increase to a peak and then reduce. The sum of the instantaneous entropies of the cumulative number densities of C_{60} $\sum S_{60}(f)$ becomes negative for $f \geq 28$.

4.7. Entropy of the emerging C_2 gas

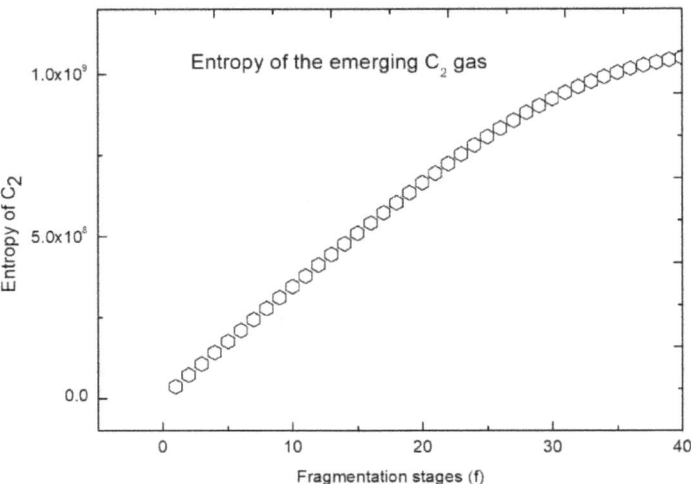

Fig. 4.8. $S_{C_2}(f)$ of the C$_2$s as a function of the fragmentation step f is positive and increasing as shown in the inset while the total entropy of all of the fullerenes and C$_2$ gas $S_{tot}(f) = \sum [S_x(f) + S_{C_2}(f)]$ is also positive and increases with f.

As discussed in the earlier sections, the nonlinear surface forces associated with the local and global curvature of large fullerenes induce instability and buckling halting their carbon accretion and growth. During the ensuing cage fragmentation sequences, the emission of a C_2 per fragmentation step relieves the local strain. C_2 emissions shrink large cages (> C_{60}) into the next smaller one. Fig. 4.8 shows the top-down sequences of fragmentations via shrinking of the large cages of the grand canonical ensemble that result in the increasing density of the associated C_2 gas. At each fragmentation step the ensembles' entropy was mapped for all the shrinking fullerenes with the positive entropy generated by the C_2s. The initial bottom-up sequence of large cage formation followed by the subsequent top-down shrinking cascades. Together, these are shown to constitute the two gases of the self-organizing, large carbon cages and the C_2s. The initial C_2-accreting stage involved these two gases and the subsequent transition towards the accumulative C_{60} gas. The stable dynamics of the self-organizing ensemble of the large fullerenes yields the C_{60} and C_2 gases.

The Stone-Wales mechanism for the conversion of fullerene bonds that lead non-icosahedral ones to the icosahedral C_{60} where out of a total of 1812 isomers has already been extensively debated in the fullerene literature [92]. We conclude that the regenerative soot there exist a large number of C_2 molecules that actively participate in the hot, fullerene forming and fragmenting environment. Such a grand canonical ensemble was treated by assuming the soot constituents to be 3D rotors. It is a unique combination of a continuous ongoing phase transformation that has a well-defined output.

Fig. 4.9 presents the collective picture of statistical mechanical entropic profiles of the transforming, fragmenting fullerenes and the emerging C_2 gas. The driving force for these transformations is the curvature-related local surface deformations of the cages. Such a dynamical description of self-organization may have important implications for understanding similar processes elsewhere in nature. In this chapter, a self-organizing grand canonical ensemble of the open and closed clusters and cages is described by using a simplified logistic map of the evolution of the C_x trajectory into C_{x-2} and C_2 via $C_x \rightarrow C_{x-2} + C_2$ (emission) and $C_{x-2} + C_2 \rightarrow C_x$ (absorption) of C_2. The simulations employed $N_{f+1} = \alpha N_f$ where $\alpha = 1/2$ for the data shown in Figs. 4.5 to 4.8. For the laser-ablated pulses the emerging soot was visualized as the gas with the constituents $\sum_{x=2}^{x=100} C_x$ that self-organizes into the C_{60} gas

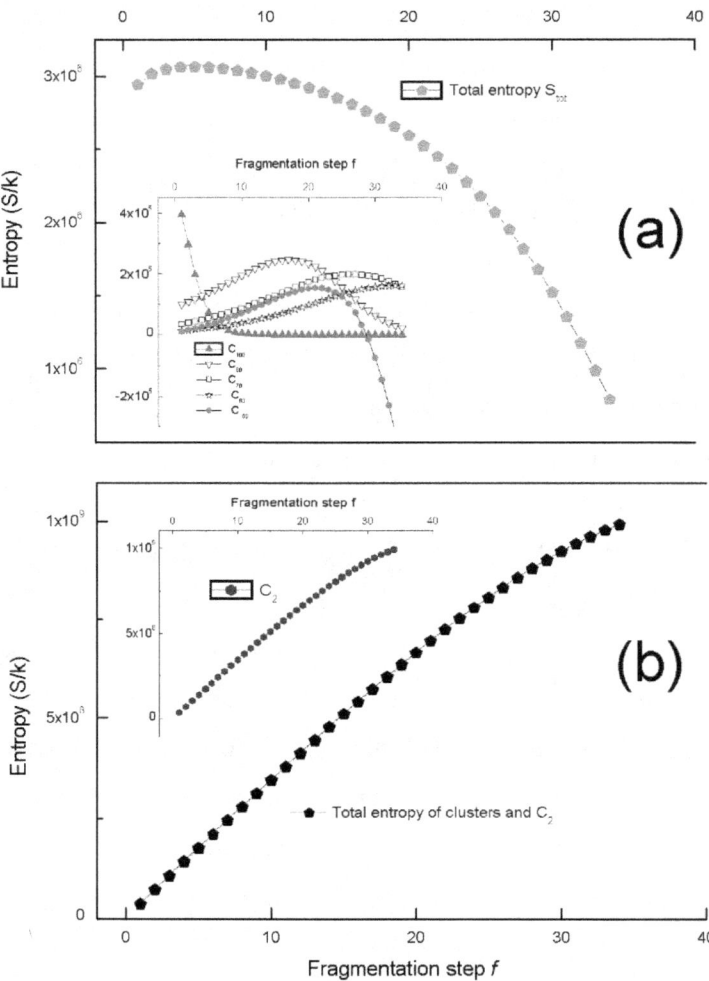

Fig. 4.9. (a) The sum of entropies of all the fullerenes $\sum S_x(f)$ from C_{60} to C_{100} shows a net decrease with increasing f. The inset shows variation of $\sum S_x(f)$ for the five selected fullerenes C_{100}, C_{80}, C_{70}, C_{62} and C_{60}. $\sum S_{60}(f)$ becomes increasingly negative for $f \geq 30$. (b) The cumulative entropy $\sum S_{C_2}(f)$ of the growing number of C_2s is positive and increasing. $S_{tot}(f) = \sum [S_x(f) + S_{C_2}(f)]$ being the total entropy of the grand canonical ensemble comprising of all the fullerenes and the associated C_2 gas is also positive and increases by each fragmentation step. Calculations were performed at 1000 K and all entropies are in units of k-the Boltzmann constant.

containing γC_{60} where $\gamma = \sum_{x=60}^{x=100} C_x$ is the sum of the isomers of all (21) fullerenes from C_{100} to C_{60}. The other constituent gas is heavily dominated by C_2s. The C_2 gas is the integral, operative and the soot-defining constituent of the dynamic, self-organizing grand canonical ensemble of the closed cages of carbon.

4.8. The regenerative soot mode

The model developed above for the pulsed laser ablation can be extended to the continuous vapor created in the regenerative soot. In that case, C_{100} can be assumed to become the source and its population does not decrease at each fragmentation stage. In this case the energy deposited in the discharge creates the continuous C vapor of the transforming, self-organizing soot. Here, one has to describe the simultaneous formation, fragmentation and re-formation of the closed cages as a continuous process. In the pulsed case scenario discussed above, C vapor condensation, results in the formation of the fullerenes according to their respective isomer population. It subsequently goes through the cage shrinking sequences of this primary ensemble. Whereas, in the continuous arc discharge, the constancy of the source density i.e. C_{100}, is maintained. C_{100} acts as the source. The cage fragmentation is treated as a serial process i.e., $C_{100} \rightarrow C_{98} \rightarrow C_{96} ... C_{62} \rightarrow C_{60}$. After the initial fragmentation stages the density of C_{62} remains constant and $\approx \sum I_x$ where x is between 62 and 100. Same is the case for the higher cages. Figure 4.10 shows the entropy directly related with the increasing density of C_2 as a function of f. $N_x(f)$ for $x = 100, 80, 70, 62$ and 60, against f show similar pattern as seen in Figs. 4.5 and 4.6. The inset in figure 4.10(a) shows the calculated entropy of the each fullerene $S_x(f)$ as a function of f. Unlike the pulsed case in figure 4.9(a), the entropies $S_{100}(f)$, $S_{80}(f)$, $S_{70}(f)$ and $S_{62}(f)$ increase with f. In the continuous discharge case, $S_{60}(f)$ initially increases, as in figure 9(a), and then decreases beyond $f \geq 20$ eventually becoming negative. This behavior of C_{60} affects the sum of the entropies of all fullerenes $\sum_{x=60}^{x=100} S_x(f)$ shown in figure 4.10(a). The sum of entropies of all fullerene cages increase up to $f \geq 20$, has a broad plateau and then decreases.

The associated entropy of the cumulative gas of C_2, S_{C_2} shows an increasing trend that is about three orders of magnitude higher than the sum of the entropies of all the rotating cages, is shown in the inset of figure 4.10(b).

The total entropy $S_{tot}(f)$ of all the rotating, shrinking cages and the C_2 liberated in the fragmentation process shows a monotonous increase in figure 4.10.

Various mechanisms for the formation of C_{60} have been actively proposed and debated ever since its discovery. This chapter has employed theoretical and simulation studies to illustrate the routes of the self-organizing carbon clusters that eventually may lead to the Buckyball and at the same time, describe the statistical mechanics of the associated negative entropy. Fullerenes-the closed cages of carbon have been shown to exist in various sizes and geometrical shapes. 28 fullerene point groups can describe all shapes and topographies of the closed cages. Icosahedral group I_h to which one isomer each of C_{20}, C_{60}, C_{80}, C_{180}, C_{240}, C_{540}... belong, has the highest symmetry. Among these, the icosahedral C_{60} is the sole survivor.

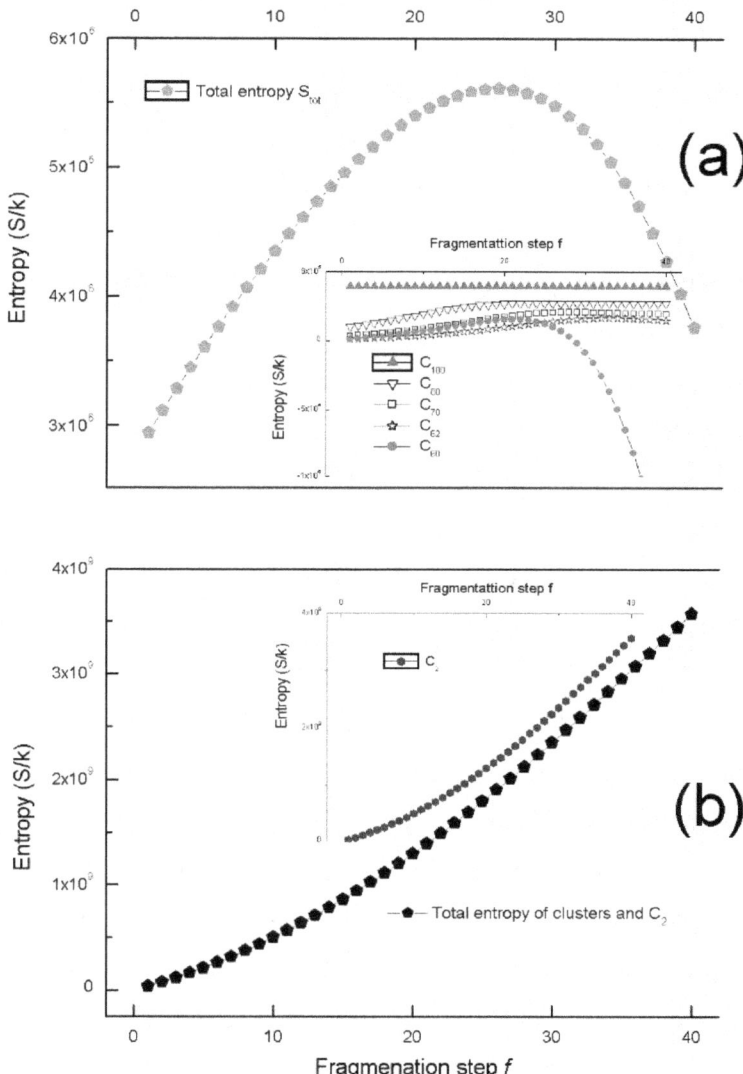

Fig. 4.10 (a) Inset: $S_x(f)$ of five selected fullerenes increase to become constant while that of C_{60} first increase to a peak around $f \geq 20$ and it becomes negative for $f > 28$. The trend of sum of entropies of all the fullerenes $\sum S_x(f)$ from C_{60} to C_{100} shows a gradual increase up to $f \approx 20$ followed by a plateau and then a net decrease of entropy with the increasing f. (b) Inset: $S_{C_2}(f)$ of the C_2s is positive and increasing. The total entropy $S_{tot}(f) = \sum S_x(f) + S_{C_2}$ of the grand canonical ensemble comprising of all the fullerenes and the associated C_2 gas shown is also positive and increases with f.

4.9. The Summary and Conclusions of Chapter 4

Simulations of the shrinking large cages have demonstrated that $C_x \rightarrow C_{x-2} + C_2$ is a plausible route for the emergence of C_{60}. Using statistical mechanical arguments, it was demonstrated in this chapter that the ensuing gas of C_2 provides the essential link for the grand canonical ensemble of the newly formed cages to deliver C_{60} as the end product. The model illustrated C_{60}'s emergence from the condensing carbon vapour. It was shown to depend upon (1) the decreasing heats of formation for larger cages, (2) exponentially increasing number of isomers for fullerenes that are larger than C_{60}, (3) large cages' buckling induced by the pentagon-related protrusions that initiate fragmentation, (4) the structural instability-induced fragmentation that shrinks large cages via $C_x \rightarrow C_{x-2} + C_2$ and (5) an evolving gas of C_2.

The simulations of the mechanisms for the provision and presence of plenty of C_2s during the formation and fragmentation processes were presented. Fullerenes were portrayed as 3D rotors with the partition functions describing ensemble's entropy as a function of the fragmentation sequence. The bottom-up formations of large cages followed by the top-down cage shrinkage were shown to be the stable, dynamical processes that lead to the C_{60} dominated, self-organizing, grand canonical ensemble of the regenerative soot.

We have extended and modified the **'order out of disorder'** paradigm into 'the self-organizing soot with the **local reduction of entropy for the C_{60}** and the **increasing, compensating entropy of the C_2 gas'**.

The Self-Organizing Soot

Chapter 5

The Degenerate Fermi Gas of Fullerenes

5.1. The σ and π electron shells of fullerenes

The energetics and structural stability of fullerenes were the subject of extensive studies since the discovery of the icosahedral 'C_{60}: Buckminsterfullerene' [2]. The foregoing chapters have illustrated these aspects. The topological arguments can provide justifications for the growth of spherical structures in the soot forming environments, and explain their comparative stability over the planar counterparts [55,93-98]. Such arguments are centered on the experimentally observed stability of the icosahedral isomer of C_{60}. Delocalization, re-hybridization and interactions of the π electronic shells with the spherical σ skeletons have been studied to provide explanations of the stability of the spherically shelled structures of carbon. The role of the strain induced by curvature in C_{60}, C_{70} and their organometallic derivatives was investigated by using the π-orbital axis vector analysis that treated the interactions of the π and σ electrons in the formation of the 3D, non-planar orbitals [99]. The strain energy of C_{60} was shown to be between 77 to 100% of the total heat of formation [85,99]. The curvature emerged as the source of strain as well as stability in fullerenes and all similar spheroidal structures. In the present chapter, the strain introduced by the structural protrusions is discussed in the context of the π electronic shells, in the open and closed C cages and the fullerenes that are $\gtrsim C_{60}$. In addition, continuing on from Chapter 3 where the continuum elasticity arguments were presented for the deformations induced by the pentagonal protrusions in the icosahedral and the non-icosahedral fullerenes, the closed cage skeletons will be addressed as the σ electronic shells. Curvature was described as the outcome of bonding in the freely growing C sheets with >20 C atoms by the addition of monomers (C_1s), diatomic (C_2s) and tri-atomic (C_3s) molecules in the carbonaceous vapor that results pentagon formation in addition to the hexagons, as described in Chapters 1, 2 and 4. The process is similar to the **bending** and **stretching** of the graphene sheets with the large, permanent deformations in the carbon's σ-bonded network in the form of pentagons along with the hexagons. Chapter 4 indicated the C-addition routes in the growth of a fullerene and the associated curvature in the simulations of the formation of C_{60} in Figures 4.1 and 4.3. The evolution of the planar σ surfaces surrounded by the π shells with the addition of pentagons and hexagons is again shown in Fig. 5.1(a) and (b). Here, the results for the formation of the closed cage shown earlier in Fig. 4.3 are depicted as the snap shots of the crucial stages of fullerene formation by the addition of C_2s. The combined effects of the σ and π electron shells of the spherically curved, open and closed C cages will be discussed to consider the conditions of the stability of fullerenes. The structures shown in Fig. 5.1(a) were presented in Figs. 4.1 and 4.3. These were obtained by the geometry optimization calculations [100]. The σ bonds are illustrated with solid sticks and the overlapping cloud as the electron cloud obtained by E_f calculations in the MNDO [101]. The pattern of growth of C_{60} presented here is for the purpose of illustration of the curvature-related properties of the π shells. It is shown to take place via an open-caged route, by C_2 insertion leading to the cage closure in the regenerative soot. Fig. 5.1(a) illustrates the role of the curvature induced electron shells of the different stages of forming of a fullerene are shown in figure 5.1(a).

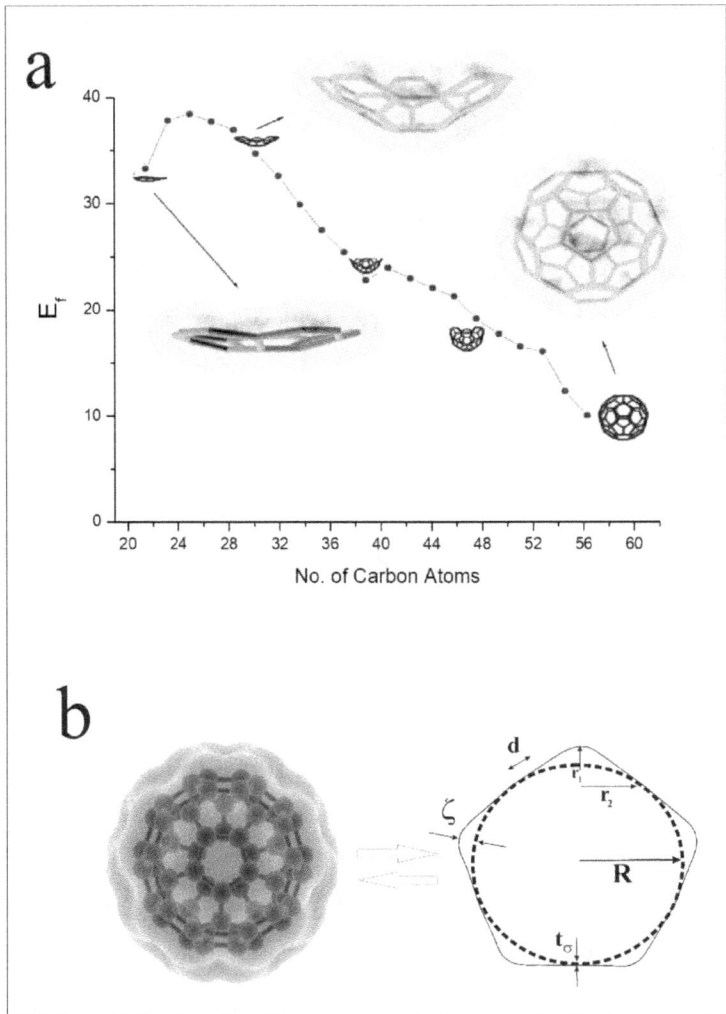

Fig. 5.1. (a) The σ skeleton surrounded by the π shells in three important formative stages of C$_{60}$ are shown. The first one is the top and side views of corannulene bowls consisting of 20 C atoms. The addition of further 10 C atoms leads to a half-cap of 30 C atoms, consisting of 6 pentagons and 10 hexagons. The addition of another 30 C atoms leads to the emergence of a C$_{60}$ shell with 12 pentagons, 20 hexagons. (b) The σ and π shells of C$_{60}$ with non-abutting pentagons are shown on the left. The equivalent σ shell, alone, is shown on the right, which is subjected to the localized pentagonal protrusions of radius r$_2$ superimposed on the global spherical shell of radius R (derived from Figs. 3.4 and 3.8).

The first and the essential stage is that of a corannulene that has just been formed; the top and side views of the corannulene with the central σ surface surrounded on both sides by the π electronic shells. This has been suggested earlier, to be one of the probable first step for introducing the straining of the graphene sheet by curvature. Topologically, it occurs whenever the sp^2-bonded C atoms assemble as a pentagon surrounded by hexagons. The top and side views of the next important stage are depicted in the half cap of C_{60} with 5 pentagons and 6 hexagons consisting of 30 C atoms. At this stage the structure is stretched by intense localized forces derived in Chapter 3 (eq. 3.8) as $f_o \approx Y t^{5/2}(\zeta^{1/2}/R)$. The final stage of the completion of the fullerene structure, where the two π shells surround the central σ structure consisting of the 12 pentagons and 20 hexagons. The figure shows clearly that as the bent and stretched graphene sheet starts to curl, the associated inner and outer π sheets have different volumes especially, near the final stages of cage closure. This effect will be more pronounced for smaller fullerenes, with abutting pentagons.

Figure 5.1 showed the significant stages of cage closure and formation of the icosahedral C_{60}. It is envisaged that all possible cages $\gtrsim C_{60}$ are formed. The self-organizational character of the closed cages in the regenerative soot, discussed in Chapter 4, provide conditions for the emergence of the grand canonical ensemble of the rotating, colliding, forming and fragmenting cages. Two basic conditions were derived that provide the route to the emergence of the icosahedral C_{60} as depicted in Fig. 5.1(a). (a) The lowering of the energies of formation E_f for the closed cages along with the increasing number of their respective isomers $I_x \propto \exp(l_0 x)$ increase the probability of formation of the cages as $P(C_x) \propto \exp(-\beta E_x + l_0 x)$, as explained in section 4.3 of Chapter 4. (b) The Stone Wales effect that provides the energetics of the bond-reversal for the abutting pentagons on the surface of a fullerene [92,102]. Together, these two become the basic conditions and the requirements for the emergence of icosahedral C_{60} in figure 5.1(a) and (b).

The nano-elastic model presented in Chapter 3 developed the theoretical arguments for the cage truncation or top-down shrinking mechanisms for the large ($>C_{60}$) cages. This model, along with the Stone Wales effect are the route for the emergence of the icosahedral C_{60} cages. In the next section, let us discuss a quantum mechanical description of the π electrons treated as the degenerate Fermi gas and investigate the two cases of the distribution of π electrons in the inner and the outer shells; (a) the equal numbers i.e. $N_\pi^{in} = N_\pi^{out} = N/2$, and (b) approximately equal densities $n_\pi^{in} \approx n_\pi^{out}$, around the σ skeleton of fullerenes with icosahedral symmetry from C_{20} to C_{1500}. The two conditions illustrate the curvature-controlling effects of the degenerate Fermi gas of π electrons. It must be noted that the entire spectrum of fullerenes are considered here to evaluate the properties of the Fermi gas of π electrons. The fullerenes belonging to various symmetry groups from the low symmetry to the highest I_h-symmetry are considered in this discussion [3]. The cages that contain the all-abutting pentagons, as in C_{20} to the non-abutting pentagons, as in I_h C_{60}. The higher fullerenes that have multiple isomers, however, the one isomer each, with I_h symmetry C_{180}, C_{240}, C_{540}, C_{960} and C_{1500}, have been used for calculations.

5.2. The degenerate Fermi gas

Let us introduce the delocalized, degenerate Fermi gas of the π electrons of the sp^2 bonded C atoms of the fullerenes. This interaction-free electron gas is the simplest approximation that neglects the interactions like the Coulomb interactions of the electrons with each other and the interaction of the electrons with the positive background of the ionic lattice. This model of the free electron gas will be employed to explain the various experimentally observed phenomena and yield physically meaningful parameters like the density of states, Fermi energies and the associated degeneracy pressures [102]. Similar, free-electron models have been used to describe the properties of the conjugated hydrocarbons based on the high mobility or delocalization of the π electrons [103-114]. The physical properties explained by using this model will include the calculations of the π electron wave functions, energy levels, dipole moments etc. In addition, by including a suitable perturbing potential, one can extend the simple interaction-free electron gas to take care of the reactivity and similar chemical properties of various π electron systems. In Schmidt's model, an aromatic system is regarded as a box filled with the Fermi gas of a finite number of π electrons [104]. This model will be extended here to the delocalized π electrons of fullerenes to learn about their role in the stability of the respective structures. An extra merit of the free-electron approximation is that one can obtain the continuous electron distributions for different fullerene structures.

For a typical fullerene shown in figure 5.2(a), the σ-bonded spherical skeleton with radius r_σ is surrounded by the inner and outer π electron shells of radii r_{in} and r_{out}. The ratio of the outer to inner π shell volumes for the fullerenes from C_{20} to C_{1500} as a function of the number of C atoms in respective fullerenes, is shown in figure 5.2(b). It is $V_\pi^{out}/V_\pi^{in} \sim 2.78$ for C_{20} and asymptotically approaches 1 for the larger ones. $V_\pi^{out}/V_\pi^{in} = 1$ is the ratio for planar graphene sheet. The resulting π electron density of the fullerene shells is also a smoothly varying parameter and the set of the two π electron clouds provides a unique system where the properties of the Fermi gas with discrete number of free electrons can be evaluated [102]. The Schrödinger equation for the degenerate π-electron gas can be considered with a Hamiltonian with kinetic energy operator alone, all other interactions are neglected and the wave function contains only the spatial part ϕ(r) takes the form

$$-\hbar^2/2m \nabla^2 \phi(r) = E\phi(r) \quad \text{eq. (5.1)},$$

where E being the one-electron energy; it has a plane wave solution

$$\phi(r) = e^{ik.r} \text{ with } E = \frac{\hbar^2 k^2}{2m} \quad \text{eq. (5.2)}.$$

To normalize ϕ(r), the electron gas can be restricted to a cube of volume V_π yielding the normalized wave function

$$\phi(r) = \frac{1}{\sqrt{V_g}} e^{ik.r} \text{ with } k_\alpha = \frac{2\pi}{L_\alpha}, \alpha = \text{x, y, z} \quad \text{eq. (5.3)}.$$

For perfect degeneracy, the fermions or the π electrons in our case, occupy all states up to a limiting momentum k_π. The number of quantum states of the translational motion of π electrons in the interval k to k +dk are $4\pi k^2 dk V_\pi/(2\pi\hbar)^3$. The total number of π electrons in respective shells and the associated Fermi momenta from zero to k_F is

$$N_\pi = \frac{V_\pi}{2\pi^2 \hbar^3} \int_0^{k_F} k^2 \, dk = V_\pi k_F^3 / 6\pi^2 \hbar^3 \quad \text{eq. (5.4)}.$$

The Fermi momentum k_F is given in terms of the electron density (N_π/V_π) by $k_F = (\frac{6\pi^2}{2})^{1/3} (N_\pi/V_\pi)^{1/3} \hbar$. The corresponding Fermi energy

$$\varepsilon_F = \frac{k_F^2}{2m} = (3\pi^2)^{2/3} \frac{\hbar^2}{2m} (N_\pi/V_\pi)^{2/3} \quad \text{eq. (5.5)}.$$

The total energy of the degenerate π electron gas is obtained by multiplying the number of states $4\pi k^2 dk V_\pi/(2\pi\hbar)^3$ by $k^2/2m$ and integrating from zero to k_F

$$E_\pi = \frac{V_\pi}{4m\pi^2\hbar^3}\int_0^{k_F} k^4\, dk = \frac{3}{10}(3\pi^2)^{\frac{2}{3}}\frac{\hbar^2}{m}\left(\frac{N_\pi}{V_\pi}\right)^{2/3} N_\pi \qquad \text{eq. (5.6)}.$$

From the equation of state, the Fermi gas obeys $P_\pi V_\pi = \frac{2}{3}E$, thus the degeneracy pressure of a finite number of π electrons in fullerene shells is

$$P_\pi = \frac{1}{5}(3\pi^2)^{2/3}\frac{\hbar^2}{2m}\left(\frac{N_\pi}{V_\pi}\right)^{5/3} \qquad \text{eq. (5.7)}.$$

The degeneracy pressure of a Fermi gas of π electrons is proportional to the 5/3 power of the electron density. In the case of the spherical fullerenes, this will imply the densities of the respective shells i.e. the inner and outer ones as n_π^{in} and n_π^{out}.

5.3. The case for $N_\pi^{in} = N_\pi^{out}$

If one assumes that the available π electrons are distributed equally in the two shells i.e. $N_\pi^{in} = N_\pi^{out} = N/2$, leading to $n_\pi^{in} = N_\pi^{in}/V_\pi^{in}$ and $n_\pi^{out} = N_\pi^{out}/V_\pi^{out}$, then the ratio of the respective electron densities n_π^{in}/n_π^{out} will be in the ratio V_π^{out}/V_π^{in}. Since $\varepsilon_\pi \propto n_\pi^{2/3}$ and $P_\pi \propto n_\pi^{5/3}$ from equations (5.5) and (5.7), large differences appear between the Fermi energies ε_π and the degeneracy pressures P_π of the two shells for all fullerenes. The differences are, however, most significant for the smaller ones and gradually reduce for the larger fullerenes. ε_π and P_π are plotted from eq. (5.5) and eq. (5.7) for the two respective π shells of fullerenes from C_{20} to C_{1500}, in figure 5.2(c-d). Fig. 5.2(c) shows that $\Delta\varepsilon_\pi (= \varepsilon_\pi^{in} - \varepsilon_\pi^{out})$ varies from 8.42 eV for C_{20} to 4.75 eV for C_{60}, and remains high even for C_{540} where $\Delta\varepsilon_\pi \approx 1.6$ eV. Similarly, in Fig. 5.2(d) the ratio $P_\pi^{in}/P_\pi^{out} \sim 5.5$ for C_{20} and reduces to about 3 for

C_{60}. These two degeneracy pressures P_π^{in} and P_π^{out}, for the larger fullerenes reduce slowly towards a common value, yet the difference $\Delta P_\pi (= P_\pi^{in} - P_\pi^{out}) \approx 1.5 \times 10^{11} Pa$ remains significant.

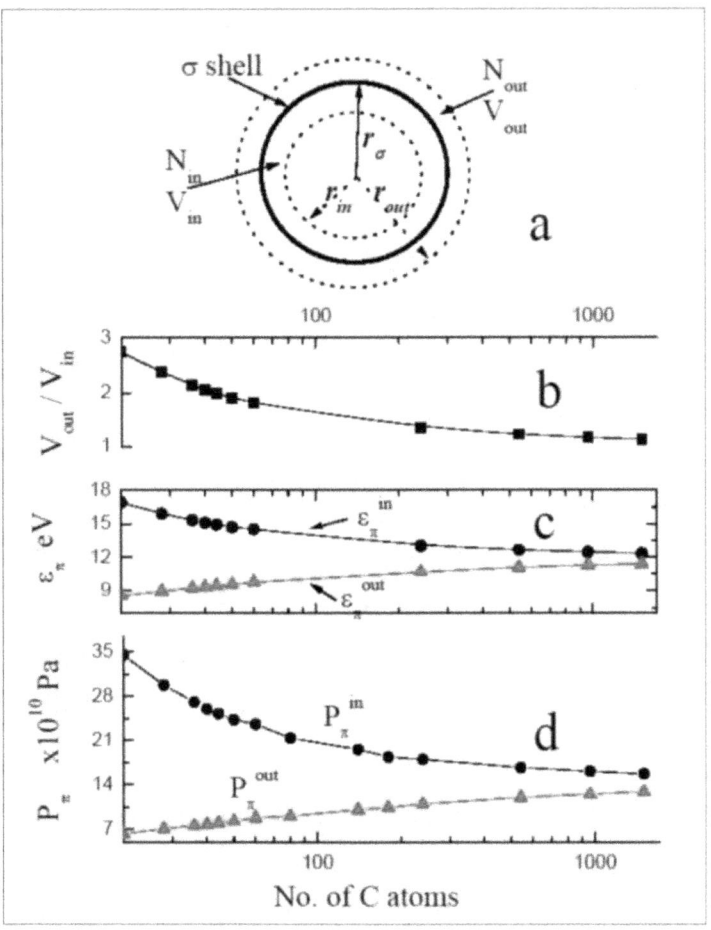

Fig. 5.2. (a) A typical fullerene is shown with its σ−bonded spherical skeleton of radius $r_σ$ surrounded by the inner and outer π electron shells of radii r_{in} and r_{out}. The number and volumes of inner and outer π electron shells N_π^{in} and N_π^{out} are also indicated. (b) The ratio V_π^{out}/V_π^{in} is plotted as a function of the number of C atoms in respective fullerenes from C_{20} to C_{1500}. (c) The Fermi energy ε_π for the inner and outer shells is shown for the same fullerene range as of (c). (d) The degeneracy pressure P_π is shown for the respective π shells.

Large differences in ε_π and P_π for the fullerenes will make these structures inherently unstable. Application of the March model to Fullerenes imply that ≈ 44% of electrons are in the inner region, with ≈ 56% of electrons in the outer region, for a typical fullerene of radius r = $6.7a_0$ ≈ 3.56 Å [98]. The case for the equal numbers of π electrons (N_π^{in} = N_π^{out}) induces structural instability. In the initial stages of the formation of the open cages, it may hinder the pentagonal protrusions. But once formed, the corannulene and the larger bowls grow with the redistribution of the π electron densities in the two shells.

5.4. The case for equal charge density $n_\pi^{in} \cong n_\pi^{out}$

In Fig. 5.2, the case of the approximately equal charge densities in the two π shells i.e. $n_\pi^{in} \cong n_\pi^{out}$, emerges indirectly when the large fullerenes are considered for which V_π^{out}/V_π^{in} approaches 1. This would be the case for the graphene sheets without any bending or the stretching effects. It also leads to the graphene sheets with the degenerate Fermi gas of the π electronic shells with $\varepsilon_\pi^{in} \cong \varepsilon_\pi^{out} \cong 12eV$ and $P_\pi^{in} \cong P_\pi^{out} \cong$ 1.4×10^{11} Pa. Equal Fermi energies induce graphene formation as the electron transitions across the σ surface are not encouraged and equal degeneracy pressures imply a uniformly squeezed, planar structure. The equality of ε_π and P_π for the two shells on its own is not a guarantee for stability when bending and stretching of the flat sheets of graphene with the condition $V_\pi^{out}/V_\pi^{in} \sim 1$ is violated by the addition of the pentagons in the growing structures.

The σ-surface generated forces due to the curvature are described below in the two models, the first by using the continuum mechanical model presented in Chapter 3 and the second is quantum mechanical

5.5. The fullerene cage stability, P_{crit} and ΔP_π

It was shown in the preceding sections that the case of an equal number of π electrons in the two shells i.e. $N_\pi^{in} = N_\pi^{out}$ introduces massive destabilizing surface forces especially for the non-symmetrical cages. The magnitude of ΔP_π, even for the open structures like the corannulene these $\gtrsim 3 \times 10^{11} Pa$. The closed cages with abutting pentagons will have localized outward directed forces per unit area. Similar effects were predicted from the nano-elastic model described in Chapter 3 and ref. [55]. Even C_{60} will have large net outward pressure, with the strain distributed equally over the entire surface. For non-icosahedral isomers, this would be the source of instability. The non-sphericity of the abutting pentagons introduces strains in fullerenes from C_{20} to C_{58} where P_π remains very large for this group of smaller fullerenes. Therefore, the fullerenes $\lesssim C_{60}$ are prone to mechanical failure that could be understood in terms of the surface forces; either due to the localized stresses of the σ surfaces, or due to the large degeneracy pressure difference of the π electrons ΔP_π across the σ surface. It must be pointed out that when considering the nano-elastic model, the eq.(3.8) $f_o \approx Y\, t^{5/2}(\zeta^{1/2}/R)$, thickness t, now, is the sum of the two π shells in the Fermi gas model.

The two models suggest that the localized protrusions and the departure from perfect symmetry is not only the source of instability in smaller fullerenes. This interpretation of the analysis given above, agrees with the March Model [98], which predicts that the discrepancy between the number of π electrons in the two shells N_π^{in} and N_π^{out} is the source of instability. The case of approximately equal charge densities in the two π shells i.e. $n_\pi^{in} \approx n_\pi^{out}$ is the physically meaningful assumption which can explain the observed stability of the perfect spheroidal cages like the C_{60} and provide answers for the instabilities of fullerenes larger or smaller than C_{60}. The equally distributed strain, as in the continuum elastic model [55] or the degeneracy pressure of the Fermi gas of π electrons via $n_\pi^{in} \approx n_\pi^{out}$ provides justifications for the emergence of the I_x C_{60} and the self-organizational character of the regenerative soot.

Figure 5.3 shows the two types of forces acting on the σ-bonded surface; ΔP_π is the net difference between the degeneracy pressures of the inner and outer π shells ($\Delta P_\pi = P_\pi^{in} - P_\pi^{out}$) and P_{crit}-the curvature-generated surface stresses from Eq. (3.9). Whereas ΔP_π is evaluated for the entire fullerene range from C_{20} to C_{1500}, the calculations of P_{crit} are done only for the icosahedral fullerenes with I and I_h symmetry. Figure 5.3(a) is the critical σ surface stress P_{crit} calculated with the thickness of the entire shell being considered i.e., $t_\sigma \approx 1.82$Å. This is the same result that was presented in our earlier paper [55]. Figure 5.3(b) the upper graph shows ΔP_π calculated under the assumption of equal number of π electrons in the two shells i.e. $N_\pi^{in} = N_\pi^{out}$ while the lower graphs is plotted for the degeneracy pressure on the σ surface by the oppositely directed pressures $P_\pi \approx 2P_\pi^{in} \approx 2P_\pi^{out}$. Here, approximately equal charge densities in the two π shells i.e. $n_\pi^{in} = n_\pi^{out}$ was assumed, therefore the two pressures equalize. ΔP_π and P_π shown in figure 5.3, are for the fullerenes starting from C_{20}, but P_{crit} is shown only for the higher fullerenes starting from C_{80} because the nano-elasticity model relates the symmetrically disposed deformations with P_{crit}, thus only the icosahedral ones and $> C_{60}$ fullerenes are candidates for such an analysis. P_{crit} for C_{60} is not shown as the pentagonal protrusions cannot be worked out in the perfectly spherical fullerene cage.

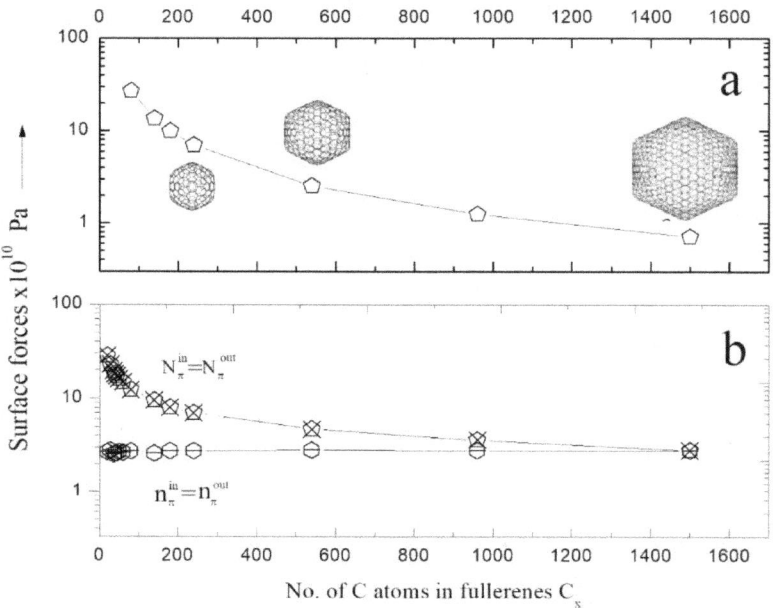

Fig. 5.3. The two types of surface forces acting on the σ-bonded structure for the fullerenes from C_{20} to C_{1500} are shown. (a) The critical σ surface stress P_{crit} calculated using Eq. (3.8) with the thickness of the two π shells being considered i.e. $t_\sigma = 1.82$ Å. (b) The upper curve is for ΔP_π - the net difference between the degeneracy pressures of the inner and outer π shells from Eq. (5.7), is shown; this is the case for an equal number of π electrons in the two shells i.e. $N_\pi^{in} = N_\pi^{out}$ as discussed in section 5.1. The lower curve has the data equal charge densities in the two π shells i.e. $n_\pi^{in} = n_\pi^{out}$. The net degeneracy pressure is $\cong 2P_\pi$.

For smaller fullerenes the magnitudes of ΔP_π and $P_{crit} \sim 10^{11}$ Pa in figures 5.3(a) and (b). These are the estimates with two extreme assumptions; one that distributes the π electrons equally in the two shells and the other where the σ − π separability is not maintained. In the latter case, the overall effect of all the electrons in the single, unified C shell is represented in the structural rigidity $D \propto t^3$, where thickness $t (\equiv t_\sigma + t_\pi)$.

In figure 5.3(a) ΔP_π is $\sim 3 \times 10^{11}$ Pa indicating the equal, large forces acting on the σ-surface due to the oppositely directed pressures of the degenerate Fermi gas of π electrons.

The trends shown in Figure 5.3 clearly indicate that the assumption of an equal number of π electrons in the two shells i.e. $N_\pi^{in} = N_\pi^{out}$ is not tenable. The large Fermi energy differences will encourage π electrons to move out from the inner shell so that the charge density in the two π shells equalizes and $n_\pi^{in} = n_\pi^{out}$. The transition from $N_\pi^{in} = N_\pi^{out}$ to $n_\pi^{in} \approx n_\pi^{out}$ occur as the crucial stage in cage closure. The π electrons are distributed in the two respective shells of closed cages in such a way that the electron densities tend to equalize i.e. $n_\pi^{in} \approx n_\pi^{out}$ making $\Delta\varepsilon_\pi$ approach zero and $\Delta P_\pi \approx P_\pi^{in} \approx P_\pi^{out}$. However, in the case of $n_\pi^{in} \approx n_\pi^{out}$ there is a tensile force from the inner π shell and a compressive from the outer π shell one acting on the entire σ-surface with $P_\pi^{in} \approx P_\pi^{out}$. Such redistribution will minimize the difference of the two Fermi energies $\Delta\varepsilon_\pi$ for the respective structures. It can be seen that the net effect of approximately equal densities $n_\pi^{in} \approx n_\pi^{out}$ is to reduce the destabilizing forces in the entire range of fullerenes irrespective of the size and symmetry. Such a range of fullerenes was seen in TOF mass spectra [2] and consistently verified by other researchers [6].

5.6. The Carbon Onions

The probabilities of formation of the larger fullerenes ($\geq C_{240}$) $P(C_x) \propto \exp(-\beta E_x + l_0 x)$ increases as compared with the smaller fullerenes ($<C_{240}$). This is due to the exponentially large number of isomers ($\propto \exp(l_0 x)$) and the reducing energies of formation (E_x). The experimental evidence suggests otherwise [2,6]. Free-standing, large fullerenes are subject to the fragmentation as discussed in the earlier

sections of the present chapter and in the last two Chapters. However, the formation of another fullerene as the outer shell may pave the way for very large fullerene formation as the fullerene-inside-fullerene configuration. This was observed in the electron microscopic studies of the soot-transforming under the energetic electrons [115-7]. These were called the Carbon Onions.

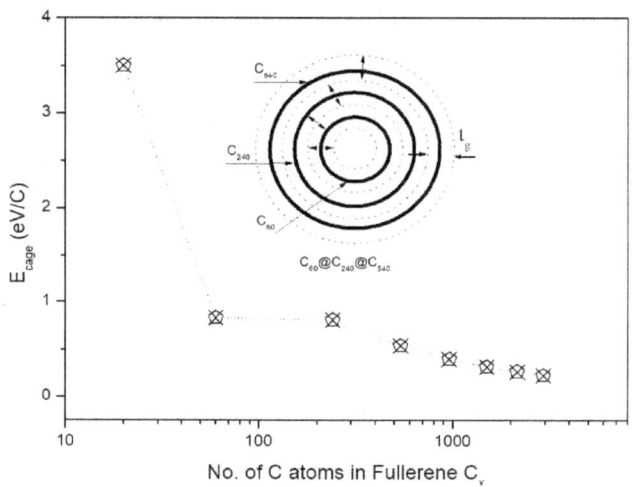

Fig. 5.4. E_{cage}/x is shown as the gradually increasing function of the Onion shell radius and the corresponding increase in the number of C atoms in each component fullerene C_x, of the C onion. The inset shown the 3-fullerene onion $C_{60}@C_{240}@C_{540}$. (from ref. [118]).

Carbon Onions are configured as $C_{in}@C_{outer}@C_{next}...$ The topological constraints are that (a) the inter-fullerene separation should be of the same order as the inter-graphene sheets in graphite, and (b) the inner and outer shells of all fullerenes must be as 'smooth' as possible, implying spheres without protrusions. In our notation this implies $C_{60}@C_{240}@C_{540}@C_{960}@C_{1500}@...$ The assumption of a thin shell the thickness t_σ becomes $\ll$ the radius of the larger fullerenes has a stabilizing effect due to the reduction of pentagonal protrusion as a result of large compressive surface forces on both sides of the central σ surface ΔP_π (= $P_\pi^{in} - P_\pi^{out}$) that tend to restore sphericity. But in such cases ΔP_π approaches the value for a graphene sheet. Although the degeneracy pressure differences may not be sufficient to compensate for the structural deformations in the free standing, $\geq C_{240}$ fullerenes, they might be responsible for the sphericity of the composite fullerene structures called the carbon onion shells, designated as $C_{60}@C_{240}@C_{540}@C_{960}@C_{1500}@$. The high resolution TEM pictures do not reveal the expected pentagonal protrusions along C_5 axis [115-117]. On the other hand, for structures ($\leq C_{240}$) the net effect of large ΔP_π may be to increase the instabilities related to structural deformations, especially for non-icosahedral fullerenes. C_{60} is the sole exception having the perfect spherical shape.

In Fig. 5.4, the cage energies per carbon atom E_{cage}/x were obtained from the pentagonal protrusion energies, discussed in Chapter 3, by using $E_p \sim Yt^2(\zeta^2/R)$ for the successive Goldberg polyhedral that represent each fullerene [3]. Although we have started from C_{20} as the first shell which has 20 trigonal protrusions, the model is however equally valid for C_{20}. Similarly, in the case of C_{60} we do not have the pentagonal protrusions as in larger fullerenes; therefore, the strain energy is evaluated from the deformation energy required to deform a shell into two flat circular disks. The results of this strain energy are shown in the same figure with C_{20} and the higher fullerenes. The energy required for producing a localized protrusion becomes larger than that required for shell's bending energy for all fullerenes $\geq C_{540}$. E_{cage}/x is a gradually increasing function of the shell radius and the corresponding increase in the number of C atoms in each fullerene C_x [118].

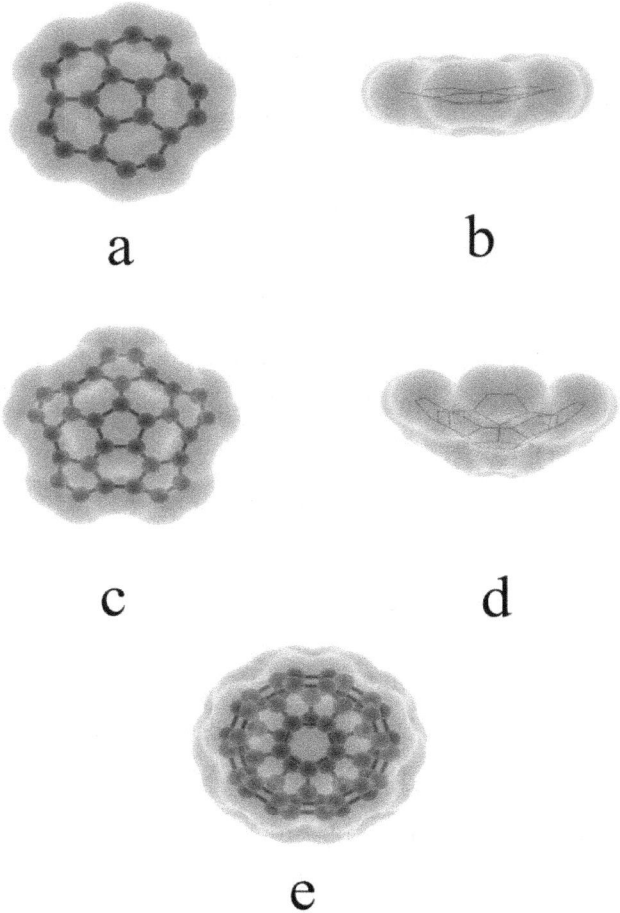

Fig. 5.5. The graphical summary of Chapter 5.

The separation of σ and π shells in three important formative stages of formation of C_{60} are shown. The top and side views of corannulene-like bowls consisting of 20 C atoms are shown as 5.5a and5.5b, the half C_{60} with 30 C atoms in 5.5c and 5.5d. The full cage with the σ and π shells in 5.5e.

The Self-Organizing Soot

Chapter 6

The End-Directed Emergence

6.1. Emerging, information-generating dynamical systems

Emergence of self-organizing dynamical systems in a variety of energy consuming, entropy generating open systems is commonly observed in nature in the form of living objects around us as the cross-linked ecological system. Over the last few decades, the emergence of a vast range of far-from equilibrium, information-generating complex natural objects and the man-made structures have been studied. A selection of the few references, relevant from the point of view of learning and understanding the phenomena of emergence in nature and some of the man-made objects in the laboratory is described in refs. [119-127]. In this chapter, we are going to develop the information-theoretic diagnostic tools to explore the emergence of the natural and laboratory-based dynamical systems by using the analogy of the inter-connected networks of information-generating and sharing constituents and components. A typical example employed to understand the emergence of such systems is shown in Figure 6.1. The main objective of Figure 6.1 is to present the concepts that will be introduced in this chapter. The figure illustrates three basic points. (a) The first deals with the complex structures that emerge due to the self-organizing, inter-connected sub-structures or the components that act as the basic ingredients of the dynamical structure. As self-organization may require energy input and the appropriate growth environment, such a structure may or may not be in equilibrium. (b) The second point relates to the generation of probability distributions that describe the evolving patterns and profiles of the emerging structures. We will show that these probability distributions provide the basis of the information-theoretic framework for the dynamic systems' emergence.

(c) The third aspect, illustrated in Fig. 6.1, relates to the core phenomenon of 'emergence' of the net outcomes. It is not always necessary for a 'new' system to emerge, rather a re-configuration or re-arrangement of the existing components and constituents may also be treated as a new setup that may generates information about the evolving, the changed or the modified configuration of the dynamical structure.

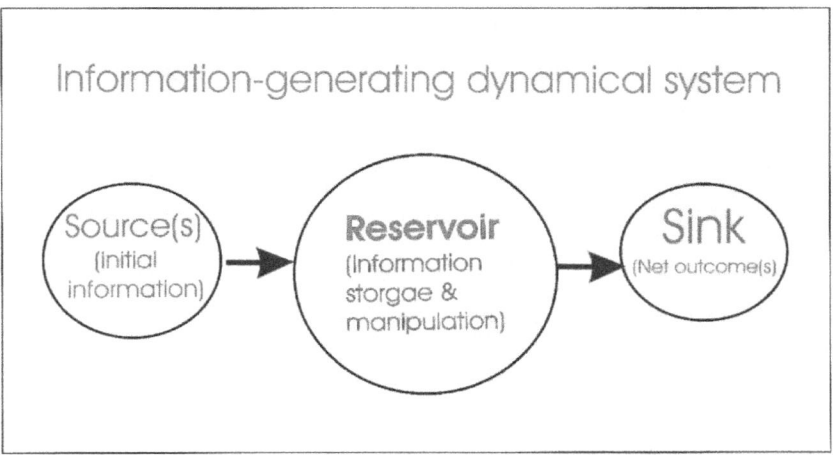

Fig. 6.1. Information-generating dynamical systems can typically consist of a Source or multiple Sources generating the initial information that is transferred to the Reservoir where it can 'reside' before onward transmission. The information received by Reservoir can be manipulated before transmission to one or multiple Sinks. Reservoir is the repository and resource of information in this three-component model of the energy-consuming, information-generating dissipative structures that may or may not be in equilibrium.

An information-theoretic Source-Reservoir-Sink (SRS) model is employed where Reservoir is analogous to Shannon's Channel between the Transmitter and Receiver [128]. The Reservoir is treated in SRS model to be the desired repository of material or the energy and the consequently generated information. The material or information received from the Source is allowed to be managed and manipulated in the Reservoir. In this respect, the Reservoir contains the undesired feature of Shannon's Channel, as a distinguishing characteristic. Here, the Reservoirs may share the manipulated and modified information with the Sink. Sink may receive, at the rates chosen by the Reservoir (i) the material output in various shapes, configurations or forms, as (ii) the signals that are representative of the processes that occurred in Reservoir and, (iii) the various byproducts that may include useable or waste material, energy, etc. Such a Sink may be the final recipient of the material or the information about the manipulative processes that occurred in the Reservoir. The interactive participation of the Reservoir with the Source and the Sink is the core of the SRS model [129]. The Reservoir may consist of one or more steps or stages of the self-organizing dissipative structures. The SRS model is developed through interconnected Boxes that may share material and information. Probability distributions are constructed that are used to calculate Shannon entropy that is referred to as Information. The fractal dimension of a dynamical system is defined which can also be considered a space filling measure that indicates the degree or the extent of spatial occupancy [130-133]. Fractal dimension is defined using Renyi's definition [134] and is also referred to and called the Information dimension in this book. Renyi defined the fractal dimension as the ratio of the total information and the logarithm of the inverse of the scale at which the information was obtained. Information dimension provides a statistical index of the complexity in such a way that illustrates how the detail in a structural pattern changes with the scale it is measured with. Another important diagnostic function called the Kullback-Leibler (KL) divergence [135] is used to compare those dynamic situations during which various probability distribution functions represent different, yet connected processes of self-organization. We designate the KL divergence as Relative Information. The Information, the fractal or Information dimension and the Relative Information will be the main diagnostic tools that are employed in this Chapter to study, investigate and quantitatively evaluate the emerging features of the far-from equilibrium, dynamic system of the regenerative soot.

6.2. Information-theoretic diagnostic tools for emerging dynamical systems

We consider the far from equilibrium, fluctuations-driven, irreversible dissipative structures that are monitored through the information generated during the spatial and temporal stages, during the emergence of dynamical system that evolves through the connected sequences Source→Reservoir→Sink as shown schematically in Figure 6.1. Information or Shannon entropy is evaluated for each constituent of the dissipative structure as it participates in the formative or the fragmentation processes. The Source is the fountainhead where the process begins with the generation and transmission of the material or the associated information. The Reservoir receives the material or the associated information generated by the Source either randomly or at pre-determined rates. The transfer rate is determined by the mutual inter-connectivity and the dynamical system-specific parameters. The Sink is the end-point of the transmitted information.

The probability paradigm is at the heart of the model. It defines the dissipative structure, the relative performance of its interactive constituents and displays the emergent characteristics of the modified or new structures. The normalized probability mass distributions for the

Source, Reservoir and the Sink are evaluated for different rates of flow and for various combinations thereof. The probability for a set of distribution steps or stages ζ is referred to as $p(\zeta)$. The uncertainty of occurrence of a certain event at step ζ is defined and measured as $\ln(1/p(\zeta))$. This function has special characteristics and has also been named 'surprise' or the Kolmogorov measure of uncertainty [136]

$$K_\zeta = \ln(1/p(\zeta)) \qquad \text{eq. (6.1).}$$

The special characteristics associated with K_ζ are related with the fact that at low $p(\zeta)$ its value increases. The sharply increasing rate of K_ζ for the decreasing probabilities, has the ' surprise' that will be illustrated later.

Information is evaluated from the sum of the product $p(\zeta)\ln(1/p(\zeta))$. The sum over all ζ of this product is the well-known Shannon entropy [128] or Information I

$$I = \sum_\zeta p(\zeta)\ln(1/p(\zeta)) \qquad \text{eq. (6.2).}$$

It must be pointed out that K_ζ and I are dependent upon the measure ζ and hence the flow rate. Different configurations of the system (SRS) will yield varying K_ζ and I corresponding to the various rates of flow and the increasing or decreasing number of the distributive stages ζ.

Information dimension is evaluated from Information $I_i \equiv \sum_\zeta p(\zeta)\ln(1/p(\zeta))$. It is the net information generated by the probability distribution $p(\zeta) \equiv p_i(\zeta)$; where $i = x, y, z$ for the respective probability distributions of the Source, Reservoir or the Sink. Information dimensional analysis based on I, for various flow rate configurations, is an indicator of the impact of the different rates on the modes of information flow of the dissipative structures. Following Renyi, Information dimension is defined as [134]

$$d_I^x = \sum_\zeta p_x(\zeta)\ln(1/p_x(\zeta))/\big(ln(1/\zeta)\big) = I/\ln(1/\epsilon) \quad \text{eq. (6.3)}.$$

In equation (3), ϵ is the number of distributive stages or the measure required to obtain Information. The fractal or Information dimension can be evaluated for the Reservoir, Source or the Sink; similar notation is used for all three components of the SRS model.

Relative Information is calculated for the emerged dynamical systems to provide a measure of the Kullback-Leibler distance between two probability measures like $p_x \equiv p_{source}(\zeta)$ and $p_y \equiv p_{sink}(\zeta)$ which are the probability distributions of the Source and the Sink, respectively. Relative Information is evaluated as [135,137]

$$D(p_x \parallel p_y) = \sum p_x(\zeta)ln(p_x(\zeta)/p_y(\zeta)) \quad \text{eq. (6.4)}.$$

6.3. Information-generating 1-Box model with the dripping faucet

The definition of the probability paradigm, used in this book for the study of dynamical systems, relies on the generation of the random as well as the controlled probability distributions. The simplest example of a box with a dripping faucet is shown in Figure 6.2. It shows the probability distributions generated by the box filled with a sharable material. In Figure 6.2(a) the material or the liquid, in the box was removed by opening a valve to randomly drain between 0 and half of the quantity remaining in the box. Four, consecutive sequences of the emptying of the box are shown. In the **inset**, the box is shown, here **1** implies the initial state of the box filled with a sharable material. The box loses its contents with a formula which ensures that a random number generator produces a number between 0 and 1/2. At each successive step, the remaining material is removed in the quantity determined by this random number generator. Four randomly generated graphs are plotted. Figure 6.2(b) has the material removal at a fixed, regular rate @1/2 to highlight a simple distributive system and to be able to define the process and the sequences of the normalized probability distribution generation at a constant rate of the material transfer.

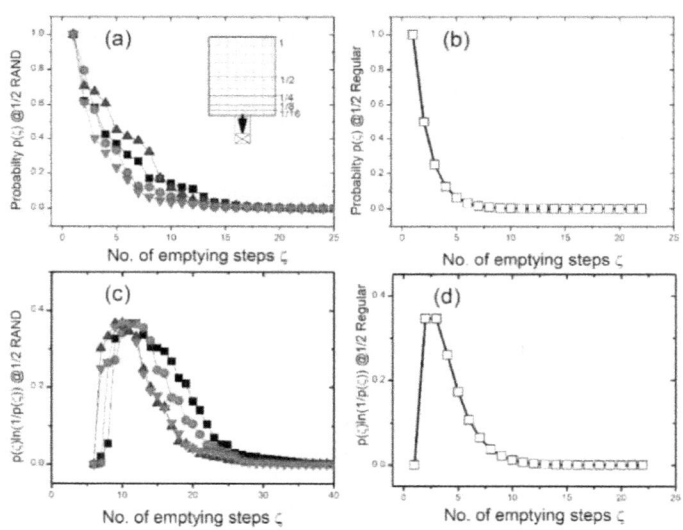

Fig. 6.2. (a) The *inset* shows four, consecutive steps of removing marbles from the box where 1 implies the box filled with a sharable material. The box loses its contents with a formula which ensures that a random number generator produces a number between 0 and 1/2. At each successive step, the remaining material is removed in the quantity determined by this random number generator. Four such, randomly generated graphs are plotted. (b) The graph shows probability p of the emptying box @1/2 at each step. This is the case of the removal of box's contents at a regular, fixed rate. (c) Instantaneous Information for the random rates is shown as four different graphs. Random rate @1/2 implies choosing a random quantity between 0 and 1/2 for the material remaining in the box for each emptying step. (d) The graph plots instantaneous information $p_x(\zeta)\ln(1/p_x(\zeta))$ as a function of the regular probability generating steps @1/2. The area under the curve is the sum of all $p_x(\zeta)\ln(1/p_x(\zeta))$ called Shannon entropy or Information written as $\sum p_x(\zeta)\ln(1/p_x(\zeta))$.

In Figure 6.2(a) and (b) the probability distribution p was generated as a function of the box-emptying steps. In 2(c) and (d), the graphs for the instantaneous information $p_x(\zeta)\ln(1/p_x(\zeta))$ against ζ are constructed by using the probability distributions $p(\zeta)$ that could either be random or regular, as in Figure 6.2(a) and (b). Sum of all $p(\zeta)\ln(1/p(\zeta))$ yields Information for the whole sequence of events, designated as $I = \sum p(\zeta)\ln(1/p(\zeta))$.

6.4. Information generating and sharing 2-Boxes model

The next step is to share the Information generated by the first box with another box. In Figure 6.3, the first box is called Source of the material to be shared with the second box- the Sink. Let us transfer a material at the rate of one eighth (@1/8) per step. It will generate Information for the Sink by receiving the transferred material. The material transfer between the two boxes is shown in Figure 6.3(a) and (b). The mechanism of generation of the Shannon entropy or Information, by the transfer of material between the two boxes in Figure 6.3 illustrates the emptying of the Source and filling of the Sink with the transferred material. Two different, yet connected, probability distributions are generated. Figure 6.3(a) is the probability $p(\zeta)$ plotted as a function of the number of emptying steps or stages ζ @1/8 per step. Figure 6.3(c) has the probability for Sink that is filling with material transferred @1/8 per step from Source. The two probabilities for the emptying and the filling of the boxes are directly related, one showing the consistent decreasing trend from the initial state [1] to [0] and the other displaying the rising pattern from state [0] towards [1]. In Figure 6.3(b), one graph is of the function $ln(1/p(\zeta))$ for the Source plotted against $p(\zeta)$. It shows a continuously increasing trend for the decreasing values of probability $p(\zeta)$. Figure 6.3(d) also has the graph of the function $ln(1/p(\zeta))$ for the filling of the Sink, with lower values of $ln(1/p(\zeta))$ as compared with those of the emptying Source. This characteristic of the function $ln(1/p(\zeta))$ is demonstrated more significantly for the smaller values of the probability due to the inverse ratio $(1/p(\zeta))$. The instantaneous information values of $p(\zeta)ln(1/p(\zeta))$ for the Source and the Sink generate different profiles in Figure 6.3(b) and 3(d), therefore their sums, or the areas under the two curves are different. The 2-Boxes model complements the information generating 1-Box in Figure 6.2 with the combined events of information-generation and transfer demonstrated in Figure 6.3.

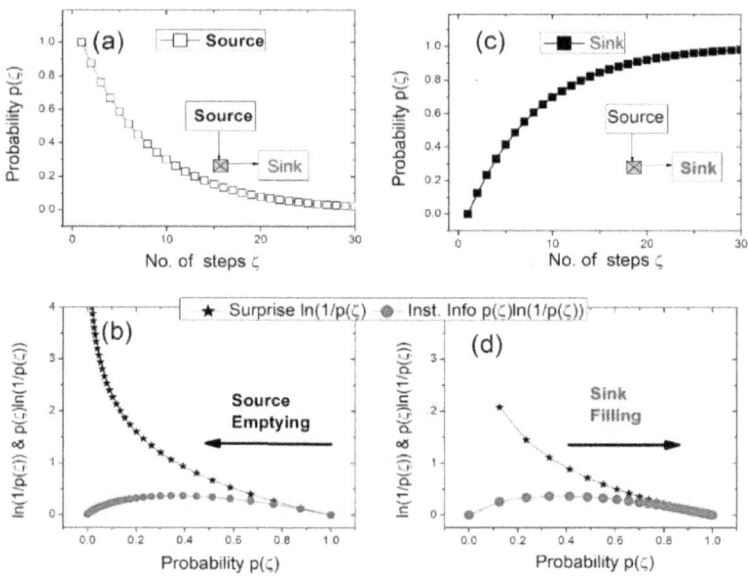

Fig. 6.3. Information transfer from Source to Sink. (a) Probability distribution $p(\zeta)$ of Source is plotted as a function of the number of emptying stages, at the regular rate of 1/8 per step. In 3(b), two functions derived from probability at each emptying step are shown, one for surprise denoted as $ln(1/p(\zeta))$ and the other for the instantaneous Information $p(\zeta)ln(1/p(\zeta))$. (c) The graph shows the corresponding probability for the material

received by Sink for each emptying step of Source. (d) The surprise or $ln(1/p(\zeta))$ is shown for the filling profile of Sink as the material is received at successive probability of each of the emptying step of Source. Instantaneous Information is also calculated and the red dots indicate their presence in (b) and (d).

6.5. Kolmogorov complexity $ln(1/p(\zeta))$

Figure 6.4 illustrates the effect of the function $ln(1/p(\zeta))$ at different rates of information transfer between the Source and the Sink. The emptying of the Source and filling of the Sink occurs at pre-determined rates. The starting probabilities for the Source and the Sink are 1 and 0 at $\zeta = 0$. The respective final probabilities for the two are ≈ 0 and ≈ 1 as $\zeta \to max$. The 2-Boxes dynamical system transforms from the initial $\{1,0\}$ state to the final $\{0,1\}$ state. However, the state $\{0,1\}$ will occur for the $\zeta \to \infty$.

Figure 6.4 is based on the iterations for the probability distributions of the 2-Box model at constant flow rate $R = (1/2)^n, n = 1 - 4$. The probability distributions and the profiles of emptying and filling vary in the inverse directions leading to the asymmetry that is introduced by $ln(1/p(\zeta))$. The Kolmogorov complexity $\ln(1/p_x(\zeta))$ is always higher for the Source at any given ζ and the instantaneous values of information $p_x(\zeta)\ln(1/p_x(\zeta))$ of each individual step ζ present different profiles for Source and Sink in Figure 6.4. The increasing values of $K_\zeta \equiv \ln(1/p_{Source}(\zeta))$ versus $p_{Source}(\zeta)$ in Figure 6.4(a), (c), (e) and (g) are due to their sharply increasing values as $p_{Source}(\zeta) \to 0$.

In the case of Source which is emptying, the largest values of K_ζ occur when the material gradually reduces $\to 0$. By comparing the values of $K_\zeta \equiv \ln(1/p_{Sink}(\zeta))$ in 6.4(b) and (d) with those in (a) and (c), the differences are strikingly clear. The instantaneous values of $p_x(\zeta)\ln(1/p_x(\zeta))$ are also shown in the eight graphs in Figure 6.4(a) to (h). When the profiles of $p_x(\zeta)\ln(1/p_x(\zeta))$ in 6.4(a) and 6.4(b) at the first few values of ζ are compared, the first value of $p_{Source}(\zeta) = 0$, and $p_{Sink}(\zeta) = 1$, the 2nd value is at $p_{Source}(\zeta) = p_{Sink}(\zeta) = 0.5$. The entropic profiles of the two boxes start at 0 and 1 and end at 1 and 0. The transformation $[1,0] \rightarrow [0,1]$ shows discontinuities in the initial values. The profiles steadily change into smooth curves only for the larger ζ.

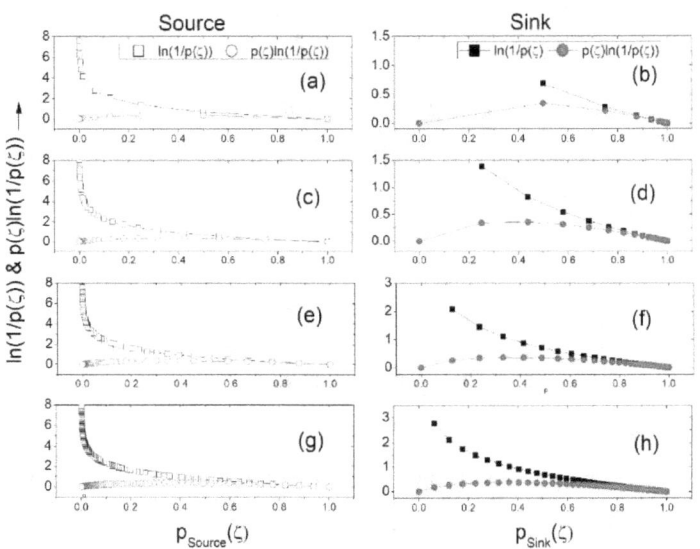

Fig. 6.4. Eight plots of the complexity $K_\zeta = \ln(1/p(\zeta))$ from eq.(1) and entropy $p(\zeta)\ln(1/p(\zeta))$ as function of $p(\zeta)$ are plotted for four rates of flow. (a) and (b) are for $R=1/2$, (c) and (d) for $R=1/4$, (e) and (f) for $R=1/8$ and (g) and (h) for $R=1/16$. The information-theoretic entropy I_x is the sum over $p(\zeta)\ln(1/p(\zeta))$ as in eq. (2).

6.6. The Source-Reservoir-Sink (SRS) model

Figure 6.5(a) shows that if a third box is added between the Source and the Sink, then a completely different flow dynamic emerges. Figure 6.5(b) shows a set of the three probability distributions for the rate of flow

$R=1/2$ from Source-to-Reservoir and $R=1/2$ from Reservoir-to-Sink. This situation resembles the case with the 2-Boxes where $R=1/2$ were shown in Fig. 6.4(a) and (b). The third box acts as the intermediate constituent in the form of a Reservoir, between the Source where the information or the original signal is created, and the Sink that receives it in the modified form after passing through Reservoir.

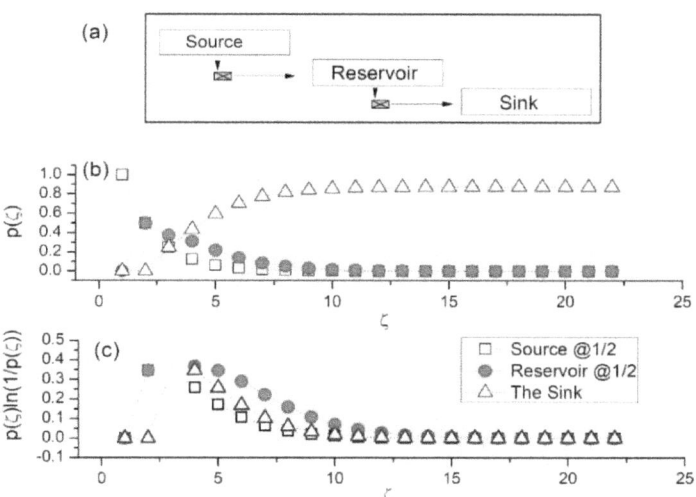

Fig. 6.5. (a) 3-Boxes schematic diagram of Source-Reservoir-Sink connected by 2 valves that regulate the flow, independently or synchronously. (b) The probabilities $p(\zeta)$ as a function of ζ are plotted for the three boxes representing Source, Reservoir and Sink. (c) Plots of $p(\zeta)\ln(1/p(\zeta))$ versus ζ shows the entropic profiles as a function of the information transfer stages ζ.

In the Shannon model [128,137], the Channel can be the source of the undesired noise, the degradation and alteration of the 'original signal' transmitted by the Transmitter. Various information-theoretic tools such as the coding and de-coding are utilized to safeguard the integrity of the 'original information'. The role of the Channel is transmission of the 'original signal' without modification. This is the core of the Information theory. In the SRS model [129] the Reservoir is utilized as the 'original signal's manipulator. The entire dynamical system (SRS) is considered as the Information generating and manipulating system. Being the repository of Information, the Reservoir's manipulative and reconstructive features can be controlled for desirable effects. The SRS model investigates the dynamical systems that may provide essential information of the emerging structural characteristics.

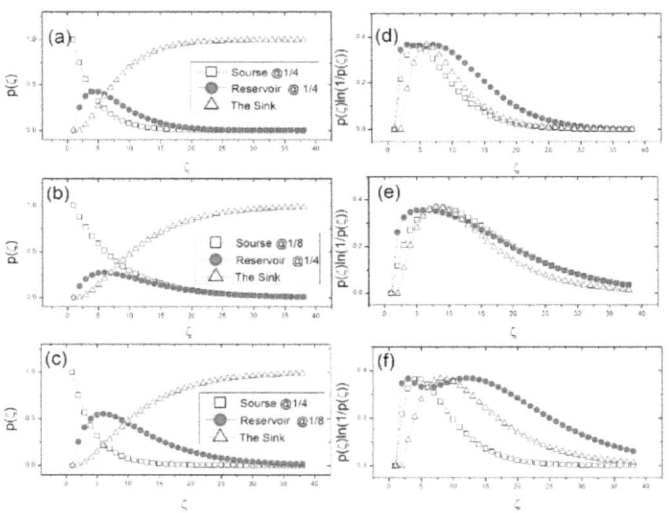

Fig. 6.6. Varying the rates of the input and discharge are shown to impact the retention capacity of Reservoir. Probabilities $p(\zeta)$ are plotted as a function of ζ for the material and/or information transfer (a) between Source to Reservoir to Sink at equal rates @1/4, (b) Source to Reservoir @1/8 and output from Reservoir @1/4. (c) Source to Reservoir @1/4 and discharge @1/8th to the Sink. The corresponding profiles of $p(\zeta)\ln(1/p(\zeta))$ are plotted in (d), (e) and (f).

Figure 6.6 has three different modes of the information-generation; (a) is the free-flow regime with equal rates of flow between the Source and Reservoir, and between the Reservoir and Sink. The mode shown in Figure 6.6(b) is the Source-friendly while the one shown in (c) is the Reservoir-friendly as the Reservoir empties into Sink at a lower rate as compared with that of the Source-to-Reservoir.

Faster rates of the arrival of material coupled with the slower rates of discharge build the capacity of the Reservoir as is evident from the comparison of 6.6(d) and (e) with (f). Figure 6.6 shows the effect of varying the rates of flow of material or information between the Source-to-Reservoir and from the Reservoir-to-Sink. Information generated by the variations in respective flows is obtained for the different combinations of flow rates.

This effect is further elaborated in the next Figure 6.7 where Information, as the sums $\sum p(\zeta)\ln(1/p(\zeta))$, is plotted for the individual constituents of the 3-Boxes at different rates of the Information transfer. In figure 6.7(a) the histograms of equal rates of flow $\{R_{Source \rightarrow Reservoir}\} = \{R_{Reservoir \rightarrow Sink}\}$ for the four different rates from 1/2:1/2 up to 1/16:1/16 are plotted. Almost equal amount of information is generated by the Source and Sink with the Reservoir capacity of Information building up with $I_{Source} \approx I_{Sink} < I_{Reservoir}$. Figure 6.7(b) has slower rate of the emptying of the Source as compared with that of Reservoir with the rates $R_{Source \rightarrow Reservoir} < R_{Reservoir \rightarrow Sink}$. In this case the information generated by the Source and Reservoir are almost similar such that $I_{Source} \approx I_{Reservoir} > I_{Sink}$. Figure 6.7(c) demonstrates that for faster arrival and slower out-flow from Reservoir that retains higher levels of material and hence, the associated information $I_{Reservoir} > I_{Sink} > I_{Source}$.

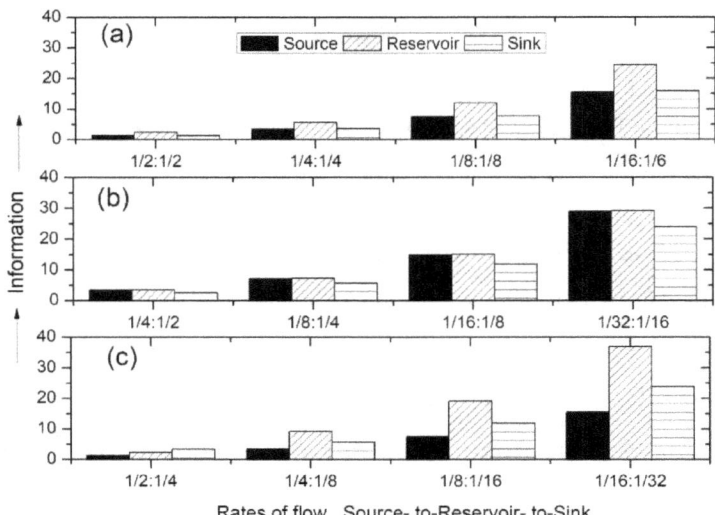

Fig. 6.7. Information as the sum $\sum p(\zeta)\ln(1/p(\zeta))$ is calculated for different sets of flow rates and shown as the three modes of information-generation and sharing. (a) The regime of equal rates of input and output allow Reservoir to build its information capacity at the expense of Source and Sink. (b) The slower input and higher output rate regime. Information generated by the Source is comparable to that of Reservoir. (c) The Reservoir-friendly mode is shown with higher rates of input and lower rates of output. The Reservoir increases the extent of Information manipulation in such a way that generates higher Information for the Sink as opposed to that of the Source.

6.7. Multiple Sources and the extended Reservoir

In the preceding section, the essential characteristics of an idealized 3-Boxes SRS model with each Box representing one of the constituents were presented. In the real physical situations, there can be more than one Sources and Reservoirs. The Sink may also not be clearly identifiable from the Reservoir. For example, even the simplest environmental example of a lake feeding on the rains and the waters from the surrounding hills and draining to a river, contains multiple Sources of water. Similarly, there can also be more than one lakes feeding a river. Investigating the cases with multiple Sources can help to determine the inter-Source or the inter-

Reservoir links. The multiple Sources configuration enhances the role and the performance of Reservoirs. Except the first Source and the final constituent- the Sink, all other intervening sources become the extended Reservoir with greater information-manipulating capabilities. Continuing with the notation of the 2- and 3-Boxes, the number of boxes is extended to 6-Boxes, sharing their contents, at a pre-determined rate with the next box. In such a case, except the first and the last one, all those in-between can bilaterally communicate to enhance or degrade the final outcome. These collectively act as the Reservoir in the following example. The model with 6-Boxes is shown in Figure 6.8.

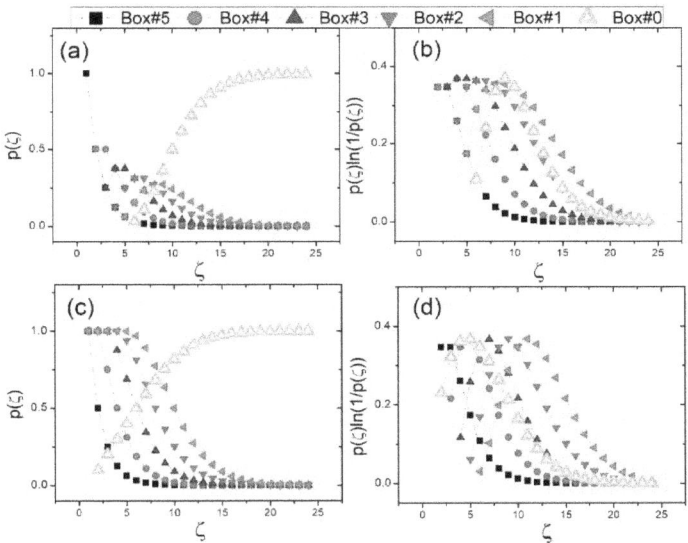

Fig. 6.8. Starting with the configuration {1,0,0,0,0,0} which implies that only the first box#5 has the entire material to be shared, all the others start with 0 material and are in the state [0], (a) the probability and (b) entropy profile for a 6-Box system with Box#5 to #0. Material transfer begins @1/2, starting with $\zeta = 0$ all of the initial material content in Box#5. (c) Probability profiles as a function of ζ when there are equal quantities in Box#5 to Box#1 in the configuration {1,1,1,1,1,0}. Box#0 has none to start with, and (d) the associated entropic profiles of the 6-Boxes. The two configurations end up with the system in state {0,0,0,0,0,1}.

The 6-Boxes setup with the interconnecting valves has similar structural arrangement as that of the 3-Boxes shown in Figures 6.5 and 6.6. The results are however, different. The 6-Boxes results are shown in Figure 6.8. All of the five boxes, Box#5 to Box#1, can initially be taken as the Sources that release their contents @1/2 to the next one. Amongst these, the set of boxes from Box#4 to Box#1 can be treated as the extended Source as well as the extended Reservoir. Box#0 is designated as the Sink. In Figure 6.8(a) and (b) all boxes, except the Box#0 share the material @1/2 per step. Figure 6.8(a) shows the probability distribution $p(\zeta)$ while in 6.8(b) the instantaneous information profile $p(\zeta)\ln(1/p(\zeta))$ versus ζ is plotted for the six boxes. The $p(\zeta)$ profile of the Boxes is such that for $\zeta \geq 8$, the Box#0 has the higher probability (or the largest quantity of material) as compared with all the other Boxes and by $\zeta \sim 12$, it is about 0.8, while for all of the others the probabilities are <0.1. The effect of Kolmogorov complexity $\ln(1/p(\zeta))$ amplifies at the successive lower values of $p(\zeta)$ that can be seen in figure 6.8(b) where the graphs for $p(\zeta)\ln(1/p(\zeta))$ of the Box#5 to #1 continue to contribute significant Information even for $\zeta > 10$ where the material transfer probability has reduced to very low levels.

The ratio of the information generated by Box#0 and the total information generated by all of the boxes is $I_0:\sum_0^5 I_b=1:6.43$. This represents the entropic cost of self-organization of the 6 interconnected boxes. The entropic cost of the dissipative process for the emergence of the last Box's information I_0 for the 6-Boxes configuration, is higher than the corresponding ratio $I_0:\sum_0^2 I_b=1:3.76$ for the 3-Boxes of the previous section. Increasing the number of boxes implies that the larger amount of information generation occurs through the interactive participation of the constituents of the self-organizing, dissipative structures resulting in the consequently the higher entropic cost for the emerging Sink as the ratio $I_0/\sum_0^5 I_b$ where $I_0 \equiv I_{Sink}$.

6.8. Information dimension d_I^x of 2-, 3-, 6- Boxes model of SRS

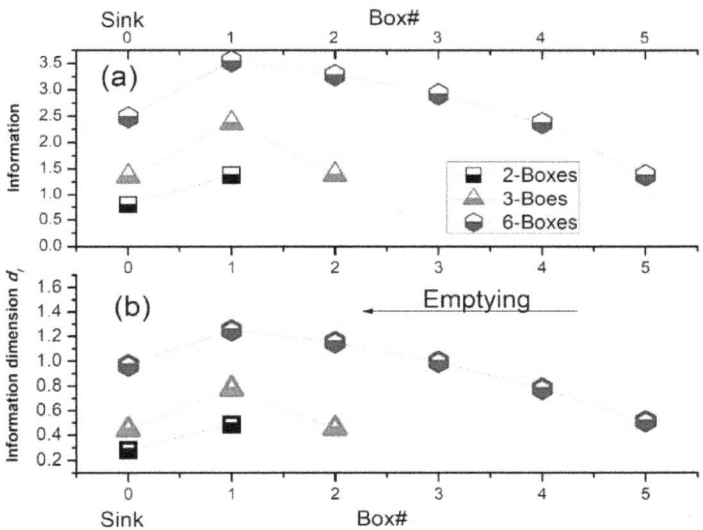

Fig. 6.9. (a) The information I_x and (b) Information dimension $d_I^{Box\#}$ of each box is shown for the 2-, 3- and 6-Boxes configurations. Box#0 is the Sink for all configurations. All boxes empty their content or share the information with the next-lower box; the emptying process is indicated by an arrow. Note that the flow rate is @1/2 for all simulations.

For the information-generating, self-organizing dissipative structures, where respective probability distributions of its components generate information $I_x \equiv \sum_\zeta p_x(\zeta)\ln(1/p_x(\zeta))$, the Information dimension was defined in eq. (6.3) as $d_I^x \cong I_x/\ln(\zeta)$. The Information-dimensional analysis for the respective information-sharing rate configurations, is shown to be an indicator of the impact of the different combinations or the modes for the constituting components of the emerging structures. The Information dimensions are obtained for the various combination of the 2-, 3- and 6-Boxes configurations. These are individually calculated and plotted for each component in Figure 6.9.

Figure 6.9(a) and (b) sum up and display the information-theoretic Shannon entropy or Information I_x and the Information dimension d_I^x of the information-generating, sharing and manipulating configurations of the 2-, 3- and 6-Boxes. These two functions identify and quantify the spatial characteristics related with the emerging, Information-sharing dissipative structures. The figure includes the three sets of the Boxes. The variations of the Information I_x and the associated Information dimensions d_I^x of the Source show gradual reduction as compared with those of the Reservoir. For the 2-Box configuration, more Information is generated by the source, hence $d_I^{Box1} > d_I^{Box0}$. In Figure 9(b), $d_I^x \equiv d_I^{Box\#}$ is plotted for each Box of the configurations discussed earlier in the preceding sections.

6.9. Relative Information $D(p_x \parallel p_y)$ of the 2- and 3-Boxes configurations

Kullback-Leibler divergence, re-designated here, as the Relative information, measures the divergence or the distance between any two probability distributions like $p_x \equiv p_{source}(\zeta)$ and $p_y \equiv p_{sink}(\zeta)$. It can be evaluated for any two components of the dynamical dissipative structures of SRS. Relative Information is defined in equation (6.4) as $D(p_x \parallel p_y) = \sum p_x(\zeta) ln(p_x(\zeta)/p_y(\zeta))$. For a truly random set of probability distributions p_x and p_y, there is asymmetry between $D(p_x \parallel p_y)$ and $D(p_y \parallel p_x)$, implying $D(p_x \parallel p_y) \neq D(p_y \parallel p_x)$. But if the two distributions are interdependent and directly linked with each other then it can be shown that $D(p_x \parallel p_y) \approx D(p_y \parallel p_x)$. This important characteristic of the relative information is employed in this book to describe the interconnectivity, or the lack of it, for the components of the dynamical system undergoing material or the information transformation-related processes.

The simple 2-Box model shown in Figures 6.3 and 6.4 displays the essential features of a connected, information-sharing system of a Source and a Sink that has the inter-dependent probability distribution functions, without any intervening Reservoir. The cumulative results for the respective information-theoretic Information I_{Source}, I_{Sink} and the associated Information dimensions d_I^{Source}, d_I^{Sink} along with the Relative Information $D(p_{Source} \parallel p_{Sink})$ are tabulated in Table 6.1. One can see that $I_{Source} > I_{Sink}$ at all rates of the generation of Information and consequently, the same is true for the corresponding Information dimensions as these are directly evaluated from I_{Source} and I_{Sink}. The difference in the values of the information I_{Source} and I_{Sink} is due to the large differences in the profiles of the Kolmogorov function $\ln(1/p_{Source}(\zeta))$ and $\ln(1/p_{Sink}(\zeta))$ when $p_{Source}(\zeta), p_{Sink}(\zeta) \to 0$. The relative information or the KL-divergence for the two probability distributions $p_{Source} \equiv p_1$ and $p_{Sink} \equiv p_0$, however, is equal $D(p_1 \parallel p_0) = D(p_0 \parallel p_1)$. This is due to the inter-dependence of the two probability distributions p_{Source} and p_{Sink}. The Kullback-Leibler distance between two random probability distributions is asymmetric but, here it is shown to be equal for the probability distributions that are directly linked with each other.

The essential features of the emerging dynamical system composed of the two inter-connected Boxes of generation and sharing of Information are presented in Table 6.1. One of the surprising result is the difference between the information-theoretic Shannon entropies or Information generated by the Source and the Sink. The same quantity of the material is exchanged between the two, yet the emerging Information differs. The detailed, step-wise analysis shows that Kolmogorov complexity has a significant effect on the generation of the net information and that this effect is most effective as $p(\zeta) \to 0$. This aspect was elaborated and discussed in Section 6.5. the Relative Information $D(p_{Source} \parallel p_{Sink})$ will be shown to emerge as a structural-process defining Information-theoretic parameter when the Information-manipulating Reservoir comes into play.

Table 6.1. The 2-Boxes. The results for I_{Source}, I_{Sink}, d_I^{Source}, d_I^{Sink} and KL Information $D(p_1 \parallel (p_0)) = D(p_1 \parallel (p_0))$ are tabulated for the six flow rates $R = (1/2)^n, n = 1 - 6$.

Flow rate Exponent n	Information $I_{Source} : I_{Sink}$	Fractal dimension $d_I^{Source} : d_I^{Sink}$	KL-distance $D(p_1 \parallel p_0)$ $\equiv D(p_0 \parallel p_1)$
1	1.39 : 0.80	0.49 : 0.28	91.13
2	3.45 : 2.17	0.97 : 0.61	165.56
3	7.45 : 4.78	1.83 : 1.18	208.49
4	15.44 : 9.96	3.20 : 2.07	458.83
5	31.38 : 20.29	5.70 : 3.69	894.79
6	63.28 : 40.92	10.19 : 6.59	1841.95

In the case of the 3-Boxes, the situation changes when an element of randomness or the disconnection of the direct route of Information exchange, is introduced with the addition of an additional box in the form of the Reservoir. Each constituent of the Source-Reservoir-Sink configuration generates Information and the associated functions that are dependent on the rates of flow between the three boxes. The calculated values of I_{Source}, $I_{reservoir}$, I_{Sink}, and the respective Information dimensions d_I^{Source}, $d_I^{Reservoir}$, d_I^{Sink} are tabulated in Table 6.2 for the regular flow rates $R = (1/2)^n, n = 1 - 4$, for the Source-Reservoir and the Reservoir-Sink.

Table 6.2. The 3-Boxes. The results for I_{Source}, $I_{reservoir}$, I_{Sink}, and Information dimensions d_I^{Source}, $d_I^{Reservoir}$, d_I^{Sink} are tabulated for the four flow rates $R = (1/2)^n, n = 1 - 4$.

Flow rate Exponent n	Information $I_{Source}: I_{Reservoir}: I_{Sink}$	Information dimension $d_I^{Source}: d_I^{Reservoir}: d_I^{Sink}$
1	1.39 : 2.37 : 1.36	0.46 : 0.78 : 0.45
2	3.45 : 5.67 : 3.52	0.96 : 1.57 : 0.97
3	7.48 : 12.03 : 7.66	1.71 : 2.75 : 1.75
4	15.46 : 24.48 : 15.87	3.16 : 5.01 : 3.25

For the 3-Boxes configuration, six Relative Information functions $D(p_x \parallel p_y)$ are constructed from the sets of the three probability distributions of the Source, the Reservoir and the Sink. The probability distributions are denoted as $p_2 \equiv p_{Source}, p_1 \equiv p_{Reservoir}$ and $p_0 \equiv p_{Sink}$. The three sets of Relative Information are; $D(p_2 \parallel p_1), D(p_1 \parallel p_2)$ between Source and Reservoir, $D(p_1 \parallel p_0), D(p_0 \parallel p_1)$ between

Reservoir and Sink and $D(p_2 \parallel p_0)$, $D(p_0 \parallel p_2)$ between Source and Sink. Table 6.3 demonstrates that the magnitude of Relative Information of the Source and Reservoir are $D(p_2 \parallel p_1) \sim D(p_1 \parallel p_2)$ for the four increasing rates of flow. This is due to the inter-dependence of the probability distributions of $p_2 \equiv p_{Source}$ and $p_1 \equiv p_{Reservoir}$. A similar situation was encountered in the 2-Box case shown in Table 6.1 where the two Relative Information were exactly equal.

Table 6.3. The 3-Boxes. Relative Information between the probability distributions for the Source, Reservoir and Sink for four flow rates $R = (1/2)^n, n = 1 - 4$. The six columns have the three sets of the two Relative Information for $p_2 \equiv p_{Source}, p_1 \equiv p_{Reservoir}$ and $p_0 \equiv p_{Sink}$.

Flow rate Exponent n	$D(p_2 \parallel p_1)$	$D(p_1 \parallel p_2)$	$D(p_1 \parallel p_0)$	$D(p_0 \parallel p_1)$	$D(p_2 \parallel p_0)$	$D(p_0 \parallel p_2)$
1	0.63	0.91	107.9	109.05	110.23	151.55
2	1.49	1.86	125.85	128.34	131.44	184.34
3	4.15	4.77	257.86	263.52	272.53	376.12
4	10.05	11.13	329.24	335.52	367.46	502.46

Relative Information of the Reservoir and the Sink in Tables 6.3 $D(p_1 \parallel p_0) \approx D(p_0 \parallel p_1)$, however, the numerical magnitude of this set is much larger than the corresponding values for the Source and Reservoir i.e., $D(p_2 \parallel p_1) \approx D(p_1 \parallel p_2)$. This is due to the fact that the twin-probability generating processes for the Reservoir's information content vary in two stages; the first stage is of receiving from the Source and in the second, it transfers part of its accumulated and the remaining contents to the Sink. Depending upon the physical constitution, conditions and the environment of the emerging dynamical system, any of the requirements would favor the manipulation of the information by Reservoir.

Reservoir, therefore, is the link pin of the dynamical system. Reservoir in SRS model receives, retains and regulates the Information. Its dominating role will be further investigated in the next chapter.

The last set of Relative Information functions between Source and the Sink, connected through the Reservoir, the two relative entropies are not equal. The last two columns of Table 6.3 clearly demonstrate that $D(p_2 \parallel p_0) < D(p_0 \parallel p_2)$ for the four flow rates. The inherent asymmetry is clearly visible. Relative Information of the Sink with respect to the Source is bound to be higher due to the nature and content of the flow towards Sink that is manipulated by the intervening Reservoir. Reservoir operates as the mediator and modulator of the Information transfer between Source and Sink. Relative Information function $D(p_x \parallel p_y)$ clearly emerges as a measure of the distance between the processes that generate the inter-connected and dependent probability distributions of the three dissipative structures constituting the dynamical system of SRS.

The Information transfer characteristics of the self-organizing, dissipative structures represented by the Information-generating and sharing SRS model are summarized in Figure 6.10. Figure 6.10(a) describes the probabilities of the transfer of information in the 2-Boxes consisting of a Source and a Sink at a regular rates of transfer @1/8. The initial state of the system's information content at $\zeta = 0$ can be represented as {1,0}, which implies that all material is in the Source and it is connected to an empty Sink. At $\zeta = 6$ the Information is shared equally between the two and the state of the combined system is now {0.5,0.5}. For the larger values of ζ, the system's final state {0,1} is asymptotically reached. The information transfer in the 2-Boxes configuration will remain same for all rates of flow, the only difference will be the position of the {0.5,0.5} state as a function of ζ.

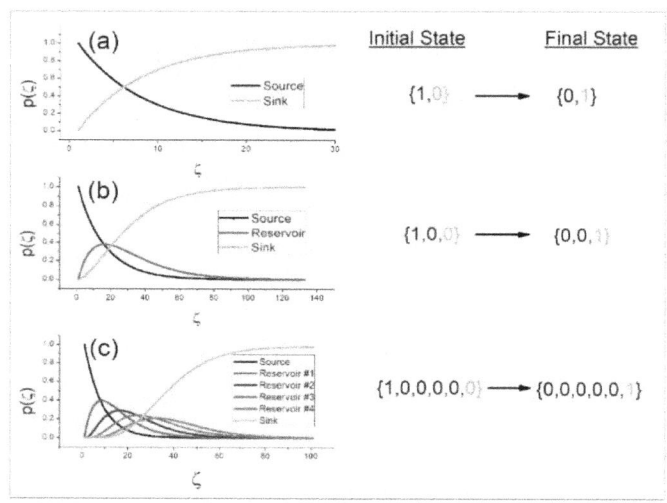

Fig. 6.10. The probabilities and the state of the system for the Information-transfer characteristics of the 2-, 3- and 6-Box configurations are shown in (a) – (c).

For the 3- and 6-Boxes configurations, the profile of the information sharing and transfer completely changes as is evident from Figures 6.10(b) and (c). The cross-over points for the probability curves for the Source, Reservoir and the Sink will change as a function of the rate of flow R and the number of the constituent boxes. The 3-Boxes information content at $\zeta = 0$ is {1,0,0} while for $\zeta = 16$ it is {0.38,0.38,0.24} and for $\zeta > 100$ it approaches {0,0,1}. Similarly in the 6-Box case, the transition from {1,0,0,0,0,0} at $\zeta = 0$ to {0,0,0,0,0,1} occurs at $\zeta > 100$. Variations in the configuration of the Boxes and the rates of flow of Information-generation and transfer will yield different final states.

6.10. Summary of the SRS Model

An information-theoretic Source-Reservoir-Sink model with the quantifiable emergent features was presented in this chapter. The SRS model has the same basic structural elements as those of the Shannon's signal Transmitter, Channel and the signal Receiver. Here, the Source symbolizes Shannon's Transmitter, Reservoir is equivalent to the noisy Channel and Sink is the Receiver of the modulated and manipulated Information. In the SRS model, the Source is considered as the fountainhead where the process begins with the generation of Information. The Reservoir receives the Information generated by the Source at random or the pre-determined rates. Depending upon the design, the style, capacity and the modes of communication with the Sink, the material and information-sharing between the Reservoir of information and the Sink occurs. The Sink, in our model, is the sole recipient of the transferred material, the associated processes and the Information from Reservoir. It represents the end-directed emergence of the dissipative dynamical system of the SRS. The information transfer rate is shown to determine the mutual inter-connectivity and the dynamical system-specific parameters and functions. The basic difference between the Source–Reservoir–Sink model [129] and Shannon's description of a Transmitter–Channel–Receiver [128] is the Reservoir's ability of manipulation, enhancement or the degradation of the Information received from the Source This is a desirable feature in the SRS model unlike the noise generated by the Channel in the Shannon model. The Reservoir can bilaterally interact and exchange Information with one or multiple Sources and Sinks. Such dissipative, Information-generating structural configurations of Source-Reservoir-Sink display self-organizational character.

Figure 6.11 is graphical summary of the chapter. It is the modified schematic representation of these SRS model-based system presented in Fig. 6.1. It includes the feedback and involution mechanisms to augment the emerging system. The involution lends an SRS configuration the ability to reach towards an end-directed emergence. The model will be extended in Chapter 7 to the emergence of the icosahedral carbon cages, the Buckyball- C_{60} out of the self-organizing soot.

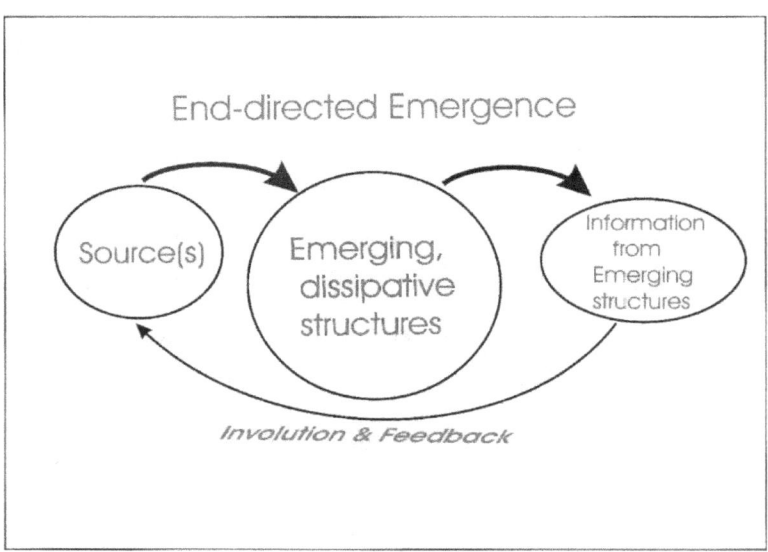

Fig. 6.11. The graphical summary of Chapter 6.

The generic diagram of the evolution of the emerging structures in dynamical dissipative environment from sub-systems or components. The manipulation of Information is shown to play the crucial role in the emergence of complex structures. A new structure may be the end-directed emergent system where an existing setup may reorganize itself by employing different dissipative mechanisms.

The Self-Organizing Soot

Chapter 7

The Dynamic Emergence

7.1. Introduction

A dynamic emergence function is defined that will be used to provide the profile of the emerging, self-organizing systems. The results of the experimental observations of the regenerative soot were discussed in the last chapters by employing the continuum elastic, Fermi gas of π electron and the SRS models. Here, the information-theoretic emergence model is extended by introducing the dynamic emergence function. The SRS model with the addition of this new function, can comprehensively describe the various stages of evolution of the dissipative structures in the irradiated SWCNTs and C_{60}. This will be done without making *apriori* assumptions about the energy-dissipating mechanisms.

Let take the example of solids irradiated by external energetic ions that can be treated as dynamical systems where the trajectories of the bifurcating recoil atoms lead to the collision cascades (CC) and localized thermal spikes (LTS). The ensuing disorder in the form of point, line and volume defects produces vacancies, sputtered atoms, dislocations and clusters of defects. Dynamics of the emergence of these defects with specific temporal and spatial profiles characterize these as the entropy-generating dissipative structures (DS). Non-equilibrium binary scattering mechanisms for CC are described by sputtering theory and Monte Carlo simulations in 3D bulk solids. The results of the C cluster sputtering from the irradiated carbon nanostructures have demonstrated the need for revisiting the theoretical framework to explain the irradiation-induced damage on the 2D surfaces of SWCNTs and the C_{60} cages.

We will demonstrate in this chapter that the dynamic emergence function provides an additional function to the SRS model to describe the emerging profiles of the fragmenting, self-organizing nanostructures of carbon.

Chapter 1 has presented the emerging features of CC and LTS in irradiated graphite and the SWCNTs. The transition of the binary atomic cascades to the collective motions of atoms as thermal spikes was considered as the self-organizational characteristics of the sp^2-bonded cylindrical sheets of graphene. Thermal models and simulations were employed to yield the atomic and cluster dissociation temperatures in SWCNTs.

The information-theoretic SRS model was presented in Chapter 6 to describe the fragmenting cages as self-organizing dissipative structures. The SRS model is firmly rooted in Shannon's information-theoretic Transmitter-Channel-Receiver paradigm where the Channel can manipulate the original signal. SRS allows the Reservoir to actively participate in generating, manipulation and transmission of the manipulated message to the Sink, which is the eventual Receiver of Reservoir-transmitted information. Here, the Reservoir operates as the emerging, information-generating and manipulating constituent, as was discussed in Chapter 6.

Following the methodology developed in the earlier chapters, the present chapter, derives Information from the experimentally determined probability distribution functions of the sputtered atoms and clusters emitted from the irradiated SWCNTs and C_{60}s as a function of irradiating ion energy. We use the data, from the extensive series of experiments on SWCNTs [138]. The C_{60} data used here, is from ref. [139].

The SRS model, in Chapter 6, was used to evaluate the Shannon entropy or Information, the Kullback-Leibler distance and the fractal dimension in the case of the fragmenting ensembles of fullerenes. A new dynamic emergence function is introduced in the SRS modular evaluations of the emerging trends of the interacting, competing, evolving and disintegrating DS of the irradiated SWCNTs and C_{60}s.

7.2. The irradiated, fragmenting SWCNTs and C_{60}-Fullerite

Probability distribution functions $p(C_x) \equiv p_x(\zeta)$ for the sputtered anions C_x are evaluated from the data of the mass spectra as a function of cesium energy $E(Cs^+)$. SWCNTs and C_{60}s were irradiated in Source of Negative Ions with Cesium Sputtering-SNICS [11]. SWCNTs of ~3 nm diameter with average length ~8 – 10 microns were compressed as targets for SNICS installed on the 2 MV Pelletron. The. Cs^+-sputtered atoms and clusters are extracted as anions. A 30° momentum analyzer delivered the mass spectra of the sputtered anions as a function of cesium energy $E(Cs^+)$. The experiments were conducted by defining the range of the Cs^+ energy $E(Cs^+)$ and choosing the appropriate scale of the incremental energy steps $\delta E(Cs^+)$. The experimental measure of dynamic emergence is $\zeta = \zeta(E(Cs^+), \delta E(Cs^+))$. Normalized yields N_{C_x} for the C_x anions with x-atoms were obtained from each mass spectrum, at a well defined Cs^+ energy $E(Cs^+)$ to determine the probabilities of emission of each cluster $p(C_x) \equiv N_{C_x}/\Sigma N_{C_x}$. The probability distribution function $p(C_x) = p_x\{E(Cs^+), \delta E(Cs^+)\} \equiv p_x(\zeta)$ represents the combined effects of the $E(Cs^+)$ and its increment $\delta E(Cs^+)$. In the experiments and the analysis thereafter, ζ is the basic measure in our experimental data with $\delta E(Cs^+)$ set at 0.1 and 0.2 keV.

C_2 emerges as the main sputtered species from all sp^2-bonded graphitic structures, as was reported in the previous chapters, during the ion-induced fragmentation experiments. The probability of emission of C_1 was shown to have an ion-energy dependence. The cascades of binary atomic collisions are likely to create single vacancies and mobile atomic defects which demonstrates the well-known ion-energy dependent probability $p(C_1) \propto E(Cs^+)$. The experiments demonstrate $E(Cs^+)$-independent cluster $p(C_{x>1})$ emissions. Chapter 1 demonstrated that clusters were the result of the emerging LTS in SWCNTs with emission probabilities $p(C_x) \approx \{exp(E_{xv}/kT_{Sub}) + 1\}^{-1}$, where E_{xv} is binding energy of an x-member vacancy at sublimation temperature T_{Sub}.

The C_2-spitting fullerenes are less likely to have the CC due to the closed nature of the cages as opposed to that of the nanotubes. The thermal nature of the ion-induced effects in the C_{60} cages will be investigated here with the help of the dynamic emergence function.

From the radiation damage perspective, the ensuing damage has two time scales $\tau_{LTS} \gg \tau_{CC}$, CC occur for fractions for ps and LTS could last for many ns. Both mechanisms have collisional relationship and share the same spatial regions. For times $> \tau_{LTS}$, the damage annealing sequences operate to minimize the localized sublimation-induced damage. A series of experiments collected the mass spectrometric data of the ion-induced fragmentations in SWCNTs and C_{60}-fullerite [138-140]. This data will be used in section and for the thermal and the SRS model calculations.

This analyses will help to identify the evolutionary trails of the emergence of the sequences of the damage. The information-theoretic SRS model will be used to evaluate the dynamic evolution of $p(C_x)$ of all of the sputtered species. The similar irradiation conditions were used for the SWCNTs and C_{60}.

The Self-Organizing Soot

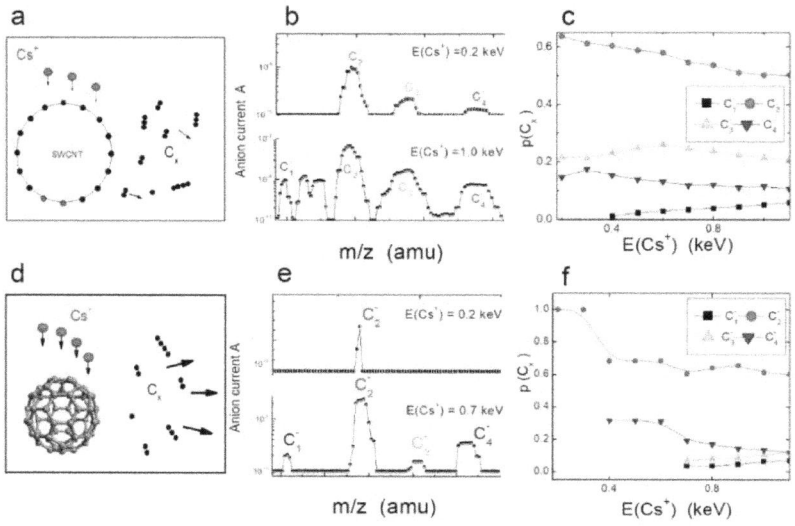

Fig. 7.1. Irradiated SWCNTs and C$_{60}$-fullerite with $E(Cs^+) = 0.2 - 1.1$ keV, $\delta E(Cs^+) = 0.1$ keV. (a) Schematics of Cs$^+$-irradiation of SWCNT and sputtering of atoms and clusters. (b) Two mass spectra of sputtered anions at $E(Cs^+) = 0.2$ and 1.0 keV are shown. Only clusters are emitted at 0.2 keV, while a low intensity C$_1$ anion is noticeable at 1.0 keV. (c) Probability distribution function $p(C_x)$ is evaluated for the four sputtered anions C_x^- from the normalized yields in each mass spectrum as a function of $E(Cs^+)$. (d) Schematic diagram of Cs$^+$-irradiation of C$_{60}$ is shown with the emitted atoms and clusters. (e) Mass spectrum at $E(Cs^+) = 0.2$ keV; only C_2^- is emitted, while all sputtered species C_1^-, C_2^-, C_3^- and C_4^- appear in the spectrum at $E(Cs^+) = 0.7$ keV. (f) $p(C_x)$ is constructed for each of the emitted species, from the mass spectra obtained as a function of $E(Cs^+)$. **Note:** in (e) the logarithmic scale is used for the C_x^- number densities to highlight the differences between the number densities of the two sets (C$_2$, C$_4$) and (C$_1$ and C$_3$), sputtered from C$_{60}$-fullerite. Data from [138,139].

Figure 7.1 presents the essential features of the experimental configuration and the results with the schematic diagrams in (a) and (d) of Cs$^+$-irradiation of a SWCNT and C$_{60}$ with the consequent emissions of atoms (C$_1$) and clusters (C$_2$, C$_3$ and C$_4$). Figure 7.1(b) shows two representative spectra from SWCNTs. At the lowest $E(Cs^+) = 0.2\ keV$, the spectrum consists of the clusters C$_2$, C$_3$ and C$_4$, while the second spectrum at 1.0 keV shows

C_1 in addition to the clusters. In 8.1(c), the normalized probabilities of emission $p(C_x)$ of the emitted species are grouped for $E(Cs^+) = 0.2 - 1.1\ keV$. Similar experimental conditions were used for irradiating the C_{60}-fullerite samples. Two mass spectra in 7.1(e) show the C_2-spitting C_{60} cages. At 0.2 keV, only C_2 is sputtered, while at $E(Cs^+) = 0.4 - 0.7\ keV$, C_2 and C_4 identify the primary routes for the cage shrinkage. The emissions of with C_1 and C_3 even as minor constituents, will be discussed later as indicators of the unstable, shrinking cages $< C_{60}$. In figure 7.1(f), the data for $p(C_x)$ as a function of $E(Cs^+)$ is plotted for the range of $E(Cs^+) = 0.1 - 1.1\ keV$. It illustrates the three distinct emerging phases of the irradiated C_{60} cages' fragmentation; from the C_2 only, emitted at $E(Cs^+) = 0.2, 0.3\ keV$, to the C_2 and C_4 emissions occur at energy range $E(Cs^+) = 0.4 - 0.7\ keV$ and all of the species C_1, C_2, C_3 and C_4 emissions at $E(Cs^+) \geq 0.7\ keV$.

The topological diversity of the static, tubular SWCNTs with a large number of C atoms ($\gg 1$) and the rotating, spherical molecules of C_{60} with the fixed number of C atoms are vividly demonstrated in their respective fragmenting patterns shown in figure 7.1. The C_x-spitting, irradiated SWCNTs transform into the damaged SWCNTs with monatomic and multi-atomic vacancies $\{Cs^+ + (SWCNT)_{pristine}\} \rightarrow \{(SWCNT)_{damaged} + C_{x \geq 1}\}$. The irradiated C_{60} cages show the sequences of the C_2 and C_4 spitting mechanisms $C_{60} \rightarrow C_{58} + C_2; C_{60} \rightarrow C_{54} + C_2 + C_4 \rightarrow \sum_{x \geq 1} C_x$ [2,3,6]. These mechanisms lead to the cases where the smaller cages with the outward, abutting-pentagonal forces, described in the Chapters 3 and 5, become unstable and may eventually disintegrate. It will be discussed in the next sections that the dynamic emergence profiles of the fragmenting SWCNTs can be described by a thermal model and the shrinking C_{60} cages will be argued with the help of a kinematical model. Conclusions drawn from the thermal models are then compared with the information-theoretic SRS model of the entropy/ information generating dissipative structures developed in Chapter 6. The additional parameter will be the dynamic emergence function.

7.3. Information-theoretic diagnostic tools for the irradiated SWCNTs and C_{60} fullerite

Irradiating ion energy $E(Cs^+)$ along with its systematic variations $\delta E(Cs^+)$ are the two basic parameters of the emerging dynamical system that is represented in Fig. 7.2 as $\{Cs^+ + (SWCNT, C_{60})\}$. It represents the initial input from the source of Cs^+ into the reservoir of sp^2-bonded networks of carbon atoms in the form of the cylindrical SWCNT or spherical C_{60} cages. This reservoir emerges as a dynamical system and a repository of the material and energy. SRS model is applied to the experimental results of the fragments sputtered from the irradiated SWCNTs and C_{60} that are shown above in Fig. 7.1. The SRS model treats such a dynamical system as entropy-generating DS where Shannon entropy or information was derived as eq. (6.1) $I_x = \sum p(C_x)\ln(1/p(C_x))$, for each of the emitted species. The two sp^2-bonded carbon nanostructures have different topologies; SWCNT is cylindrical while C_{60} is a spherical nano-cage. The other significant difference is the nature of Cs^+-target interactions. SWCNT presents a static, hexagon-based, nano-structured cylindrical target where the energy dissipation creates cascades of the binary recoiling atoms leading to the thermal spikes that originate from the enhanced localized vibrational motion of the atoms of hexagons of SWCNTs. In the case of C_{60}, a rotating, vibrating molecule is irradiated. Even in the condensed form of C_{60}-fullerite, where the translational motion is restricted, however each of the C_{60} molecule continues to rotate around a randomly oriented axis.

Continuing on from the Chapter 6, the emerging DS can be diagnosed with Renyi's definition of fractal dimension $d_I^x = I_x/\ln(1/\zeta)$, as eq. (6.2). Here, ζ is the measure of the energy incremental scale. d_I^x is calculated for every C_x from its information I_x. It will be shown that d_I^x emerges as a DS-defining function. It can also be used to distinguish the linear versus the nonlinear physical mechanisms.

Similarly, in eq. (6.3) the Kullback-Leibler (KL) distance was derived; it will be used for any two probability distributions $p_x(\zeta)$ and $p_y(\zeta)$ of the various sputtered species, as $D(p_x \parallel p_y) = \sum p_x(\zeta) \ln(p_x(\zeta)/p_y(\zeta))$. It is a measure that will be used to confirm the nature of the irradiation-induced, diverse physical processes. For example, if the monatomic species' sputtering is due to a linear CC while the C_2 and C_3 are considered the flag-bearers of the nonlinear thermal spike (LTS) then $D(C_2 \parallel C_1) > D(C_2 \parallel C_3)$. This would imply that Kullback-Leibler distance between the probability distributions of C_2 and C_1 will be larger as compared with the distance between $p(C_2)$ and $p(C_3)$ that have the same origin i.e., the LTS. KL distance identifies that two distinctly different physical mechanisms operate for the sputtering and emissions of C_2 and C_1 from the irradiated SWCNTs. Similarly, the shrinking stages of the C_{60} cages will be identified.

The experimental data on the probabilities of emission of the atomic and cluster constituents from irradiated SWCNTs and C_{60}-fullerite as a function of the operating parameter $E(Cs^+)$ can be used to obtain all of the relevant functions for the SRS model including the dynamic emergence function $\varepsilon_x(\zeta)$, that will be defined in the next section.

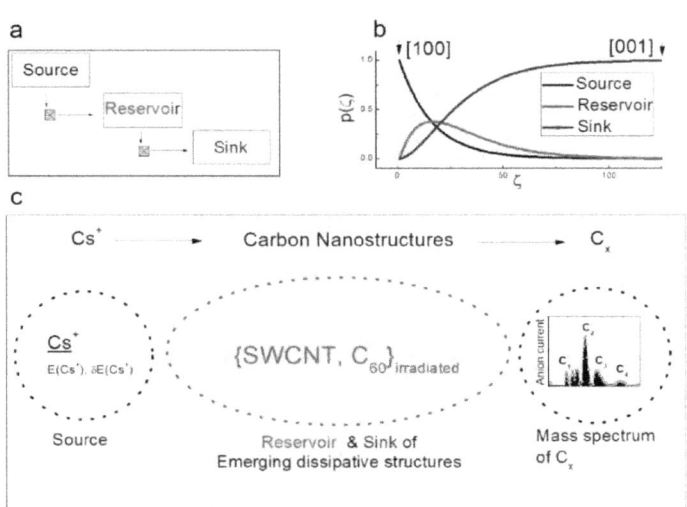

Fig. 7.2. SRS-model applied to the irradiated-SWCNTs and C$_{60}$. (a) Source-Reservoir-Sink (SRS) model is shown as information generating, manipulating and sharing Boxes. (b) Probability distribution function $p(\zeta)$ is constructed for each Box at a given flow-rate of information between the boxes. $p(\zeta)$ is plotted as a function of ζ which is defined as the number of information-sharing stages, at a given flow-rate. The initial state of SRS is pointed with arrow as [100]. It transforms to [001] for large ζ, i.e., when the information transfer from the Reservoir to the Sink has completed. (c) The region shown as dotted ellipse is where information-sharing, generating and transformation is happening. The physical representation of SRS is shown where Cs$^+$ (representing Source) irradiates carbon nanostructures with the consequent emission of C_x^-. Reservoir to Sink transformation is represented as $\{Cs^+ + (SWCNT, C_{60})_{pristine}\} \rightarrow \{(SWCNT, C_{60})_{damaged} + C_{x\geq1}\}$. Mass spectrum of the emitted C_x^- carries signatures of the emerging dissipative structures that yield experimental probabilities $p(C_x)$.

Fig. 7.2. shows the information generation by the Source, manipulated by the Reservoir before being transmitted to the Sink. Figure 7.2(a) and (b) are the modular description of the SRS model of Chapter 6, applied to the Cs-irradiated nanostructures. The model will be employed in Chapter 8 to describe the emergence of the icosahedral C$_{60}$ out of the ensembles of the self-organizing fullerenes. Here, it is employed, with an additional function defined as dynamic emergence, to investigate the emerging trends of the fragmenting, irradiated SWCNTs and C_{60}s. The results of the SRS model will be compared with those obtained by evaluating these two irradiated nanostructures by the two equivalent thermal models. The experimentally determined probabilities $p(C_x)$ of the sputtered carbon atoms and clusters $(C_x; x = 1\ to\ 4)$ will be used in the calculations for all functions of the SRS model.

7.4. The Dynamic Emergence function $\varepsilon_x(\zeta)$

The dynamic emergence function $\varepsilon_x(\zeta)$ is defined and introduced in this chapter, evaluated for all of the sputtered species C_x as a function of ζ, to provide the relative profiles of all constituents of the emerging dynamical systems. Dynamic emergence function is defined as

$$\varepsilon_x(\zeta) = p_x(\zeta)\ln(p_x(\zeta))/\sum_x p_x(\zeta)\ln(p_x(\zeta)) \quad \text{eq. (7.1)}.$$

The function is the ratio of the instantaneous information of the particular object and the collective information of all objects at the chosen step or instant ζ_0. For example, in the case of sputtering of 4 constituents C_1, C_2, C_3 and C_4, dynamic emergence function $\varepsilon_2(\zeta)$ will be the set of values of the ratio of the instantaneous information of C_2 with respect to the information generated by all $\sum_{x=1-4} p_x(\zeta)\ln(p_x(\zeta))$. It is written and calculated from the experimental data as

$$\varepsilon_2(\zeta_0) = (p_2(\zeta_0)\ln(p_2(\zeta_0))/\{\sum_{x=1-4} p_x(\zeta_0)\ln(p_x(\zeta_0))\} \quad \text{eq. (7.2)}.$$

It will be shown that $\varepsilon_x(\zeta)$ and its derivative $d\varepsilon_x/d\zeta$ can be used to map the evolution of the diverse nature of the ion-induced structural transformations, fragmentations etc., as a function of ζ. The set of the four SRS model's diagnostic tools I_x, d_f^x, $D(p_x \parallel p_y)$ and $\varepsilon_x(\zeta)$ will provide the coherent trends of the emerging and evolving DS without *apriori* assumptions about the physical mechanisms.

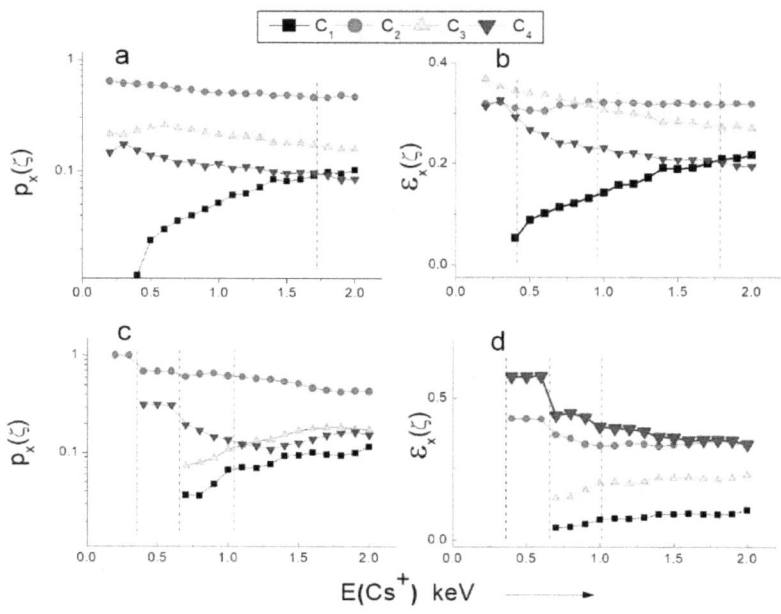

Fig. 7.3. The probability distributions $p_x(\zeta)$ and the dynamic emergence $\varepsilon_x(\zeta)$ as a function of ζ. (a) **SWCNTs:** $p_x(\zeta)$ as a function of ζ for C_1^-, C_2^-, C_3^- and C_4^- are plotted with Cs^+ in the energy range 0.2-2.0 keV with $\delta E(Cs^+) = 0.1\ keV$. (b) $\varepsilon_x(\zeta)$ versus ζ for SWCNTs show the increasing dynamic emergence of monatomic C_1 with $d\varepsilon_1/d\zeta > 1$, C_2 to C_4 have steady ε_x profiles. (c) **C60-fullerite:** $p_x(\zeta)$ for C_1^-, C_2^-, C_3^- and C_4^- for the same energy range as (a), only C_2^- emitted below 0.4 keV, C_2^- and C_4^- up to 0.6 keV. C_1^- and C_3^- emerge at 0.7 keV. (d) $\varepsilon_x(\zeta)$ versus ζ for C60-fullerite has distinctive emergent behavior of C_4^-.

Figures 7.3(a) and (c) have the two sets of experimental data of for $p_x(\zeta)$ as a function of ζ, one for the SWCNTs and the other for C_{60} [31,139, 140]. All data points are plotted as a function of ζ for each value of $E(Cs^+)$ at the chosen scale $\delta E(Cs^+) \equiv \zeta$. The same energy range $E(Cs^+) = 0.2 - 2.0\ keV$ and $\zeta = \delta E(Cs^+) = 0.1\ keV$ is used in Fig. 8.3(a) and (c) where $p_x(\zeta)$ versus ζ are plotted for the SWCNTs and C_{60}-fullertie. Dynamic emergence $\varepsilon_x(\zeta)$ for C_1-C_4 derived from data in 7.3(a) and (c) are plotted in Fig. 7.3(b) and (d), respectively.

C_2 can be seen as the most abundant sputtered component, consistently at all $E(Cs^+)$ from the irradiated SWCNTs and C_{60}-fullerite. The fractal dimension $d_f^2 \sim 2$ for C_2 for the fragmenting SWCNTs and C_{60}s in Fig. 7.3.

(A) SWCNTs: The $p_x(\zeta)$ and the dynamic emergence function $\varepsilon_x(\zeta)$ for the sputtered species $C_x; x \geq 1$ from the irradiated SWCNTs are shown in Fig. 7.3(a) and (b). There are at least four noticeable features of the data for the probabilities $p_x(\zeta)$ and the corresponding $\varepsilon_x(\zeta)$. (1) The $p_x(\zeta)$ for the cluster emission are in the sequence $p_2(\zeta) > p_3(\zeta) > p_4(\zeta)$. (2) $p_1(\zeta)$ displays $E(Cs^+)$-dependence $p_x(\zeta) \propto E(Cs^+)^m$, where the exponent m is a function of physical characteristics of the ion-target combinations [17, 25]. Its dynamic emergence function $\varepsilon_1(\zeta)$ displays consistently increasing trend with the increasing $E(Cs^+)$ and its derivative $d\varepsilon_1(\zeta)/d\zeta > 0$ (3) The dynamic emergence function for C_2 is constant for the entire range of $E(Cs^+)$, implying $d\varepsilon_2(\zeta)/d\zeta \approx 0$. C_3 shows a gradual decline while the $\varepsilon_4(\zeta)$ shows a sharp reduction. (4) In Fig. 8.3(b), clusters $d\varepsilon_{2,3}(\zeta)/d\zeta \sim 0$ and a negative trend for C_4 suggests its dissociation via $C_4 \rightarrow C_3 + C_1$ and $C_4 \rightarrow 2C_2$. The rates of dissociation will be presented in a later section. Emission of C_1 is also associated with CC that lead to LTS.

The binary collision-based cascades are linear dissipative structures with calculated fractal dimension of C_1 in Fig. 7.3(a) is $d_f^1 \sim 1$. Therefore, the CC contribution to C_1 must be significant fraction as compared with the fraction coming from the dissociating C_4. Fractal dimensions of C_2 and C_3 are $d_f^2 \sim d_f^3 \sim 2$, evaluated from their respective $p_x(\zeta)$ in Fig. 7.3(a). These describe the LTS as space-filling multifractal [133].

(B) C_{60}-fullerite: C_{60} fullerite is the condensed state with face-centered cubic structure. The $p_x(\zeta)$ and the dynamic emergence function $\varepsilon_x(\zeta)$ for the sputtered species $C_x; x \geq 1$ from the irradiated C_{60} fullerite are shown in Fig. 8.3(c) and (d). Irradiated with Cs^+ ions in the similar energy range $E(Cs^+) = 0.2 - 2.0\ keV$, that was used for SWCNT irradiations described above. However, in this case, it leads to cage shrinking of the pristine C_{60} with C_2 and C_4 emissions $cage \rightarrow cage + C_x; x = 2,4$. The fragmentations reduce the original numbers of C_{60} and generate successively increasing number densities of C_{58} and C_{56}.

The $p_x(\zeta)$ versus $E(Cs^+)$ data shown in Fig. 7.3(c) is divided into 4 parts by the vertical dotted lines. (1) The first is the C_2-only emission regime at $E(Cs^+) = 0.2\ and\ 0.3\ keV$. Cages shrink via $C_{60} \rightarrow C_{58} + C_2$. (2) This stage is followed by the C_2 and C_4 emissions in the range $E(Cs^+) = 0.4 - 0.7\ keV$. This is the regime of the two probable cage shrinking events $C_{60} \rightarrow C_{58} + C_2$ and $C_{60} \rightarrow C_{56} + C_4$. The ratio $p_2(\zeta):p_4(\zeta)$ is 7:3. (3) All fragments' regime occurs with $E(Cs^+) > 0.7\ keV$. Here all possible routes of fragmentation of the irradiated cages are followed, with the emission of C_1, C_2, C_3 and C_4. (4) C_2 maintains a steady and almost constant probability of emission between $E(Cs^+) = 0.4 - 2.0\ keV$ i.e. after the initial phase when only C_2 is emitted ($E(Cs^+) < 0.4\ keV$). (5) For $E(Cs^+) \gtrsim 1\ keV$, the number densities of C_3 becomes comparable with C_4's. Monatomic C_1 is very steadily rising, not as sharply as in Fig. 7.3(a). Here, in the spherical cage of a limited number of collision partners, the binary atomic cascades may not be operating, as these do in the cylindrical cages of nanotubes with large number of C atoms.

Fig. 7.3(d) confirms the above mentioned mechanisms of the C_{60} cage shrinkage. It plots the dynamic emergence function $\varepsilon_x(\zeta)$ as a function of $E(Cs^+)$. It is the ratio of the relative, instantaneous information of any one constituent with respect to the sum of all constituents' information. For example, for C_2 emission from the irradiated cage, its normalized probability with the all-inclusive probability, its dynamic emergence function

$$\varepsilon_2(\zeta_0) = (p_2(\zeta_0))\ln(p_2(\zeta_0))/\{\Sigma_{x=1-4} p_x(\zeta_0)\ln(p_x(\zeta_0))\}.$$

These are plotted in Fig. 8.3(d) for C_2 and all other sputtered constituents. the notable differences of the profiles of $\varepsilon_x(\zeta)$ as compared with those of $p_x(\zeta)$ are: (1) the lower probability generating C_4 now has higher $\varepsilon_4(\zeta) > \varepsilon_2(\zeta)$ for the entire energy range from $E(Cs^+) = 0.4\ keV$ to $E(Cs^+) = 2.0\ keV$. (2) The inclusion of C_1 and C_3 for $E(Cs^+) \gtrsim 0.7\ keV$ indicates that the shrinking cages are unstable and likely to fragment. The continued irradiations lead to the shrinking cages with reducing number of hexagons. A $C_2 \equiv 1\ hexagon$. Icosahedral C_{60} with twenty hexagons has twelve non-abutting pentagons. The reducing number of hexagons and increasing numbers of abutting pentagons induce non-isotropic distribution of the cages' strain as discussed in Chapters 3 and 5. The strained cages are likely to explode when the shrinking $C_{60} - nC_2 \rightarrow C_{32}$. C_{32} is claimed to be the last of the stable fullerenes. For cages $<C_{32}$, explosion is most likely into fragments $C_{30} \rightarrow \Sigma_1^6 C_x$. The relative emission probabilities of these fragments, at the explosion stage, are according to their binding energies in the pre-explosion cages. Calculated d_I^x of clusters $d_I^2 \sim d_I^3 \sim d_I^4 \sim 2$,

similar to SWCNTs in Fig. 8.3(a). C$_1$'s fractal dimension remains $d_f^1 \sim 1$ which indicates the linear character of the mechanism responsible for its emission.

Table 7.1. Comparison of the physical parameters derived from thermal and kinematical models with the diagnostic tools derived from Information-theoretic model.

Cs$^+$-irradiated Carbon Nanostructure	Thermal and Kinematical Models T_{Sub}	SRS Model d_f^x and $d\varepsilon_x(\zeta)/d\zeta$
SWCNT	$p_x = \{exp(E_{xv}/kT_S) + 1\}^{-1}$ $T_{sub} \cong 4000 \pm 250\ K$	$d_f^2 \sim d_f^3 \sim 2$ $d_f^1 \sim 1$ $d\varepsilon_1(\zeta)/d\zeta > 1$
C$_{60}$-Fullerite	$K_x = [Z(C_{60-x})Z(C_x)/Z(C_{60})]\, e^{-\frac{E(C_x)}{kT}}$ $T_{sub} \cong 2200 \pm 200\ K$	$d_f^2 \sim d_f^4 \sim 2$ $d_f^1 \sim 1$ $d\varepsilon_{2,4}(\zeta)/d\zeta \sim 0$

7.5. Thermal and the SRS models of the fragmenting SWCNTs and C_{60} cages

Two thermal models are used here. The first one was introduced in Chapter 1 for the static targets like SWCNTs. Localized thermal spikes were shown to be responsible for cluster sputtering. The rolled graphene sheets of SWCNTs were characterized in the irradiated-SWCNTs in figure 7.3 by the fractal dimensions of the clusters sputtered $d_f^2 \sim d_f^3 \sim d_f^4 \sim 2$. The 2-dimensional localized regions are thus identified where the temperature on the surfaces $\sim T_{sub}$. Equation (1.1) utilizes the probability of emission of a cluster C_x with energy of formation E_{xv} at temperature T_{sub} $p(C_x) \approx (exp(E_{xv}/T_{sub}) + 1)^{-1}$. The experimentally measured ratios of any of the two clusters $p(C_x)$ and $p(C_y)$ was used to yield the sublimation temperature in eq. (1.2) of Chapter 1 as

$$T_{Sub} \cong [(E_{xv} - E_{yv})/k)][\ln(p(C_y)/(p(C_x))]^{-1},$$ here, $p(C_x)$ and $p(C_y)$ are not dependent on $E(Cs^+)$ but on T_{sub}. This was demonstrated in Fig. 7.3(a) for the $p(C_x)$ spectra as a function of $E(Cs^+)$. Sublimation temperature calculated from the ratios of two of the clusters are shown in Fig. 7.4(a). The ratio of the probabilities $p(C_2)/p(C_3)$ sputtered from the LTS regions on the surface of the irradiated-SWCNT and the kinematical ratio K_2/K_3 described below (eq. (7.2)) are both plotted as function of temperature. The coincident ratio of $p(C_2)/p(C_3) \sim 3$ in figure 8.3(a) with K_2/K_3 yields $T_{Sub} \sim 4000 \pm 250$ K.

For the rotating spherical cages, the second thermal model employs the statistical thermodynamically determined partition functions for the chemical equilibria of the reactions $C_x \rightarrow C_{x-y} + C_y$. For the irradiated fullerite, the partition functions and binding energies for the rate equations are applicable to the C_{60} cages [141]. The irradiated fullerene cages of C_{60} and the shrinking cages of C_{58}, C_{56}..., are in higher vibrational and rotational excited states. Rates of fragmentation K_x of C_{60} via reactions $C_{60} \rightarrow C_{60-x} + C_x$ are calculated by evaluating the partition functions $Z(C_x)$ of the fragmenting cages and the fragments as

$$K_x = \frac{Z(C_{60-x})Z(C_x)}{Z(C_{60})} e^{-\frac{E(C_x)}{kT}} \qquad \text{eq. (7.2)}.$$

Where the partition functions of translation $z_{tr}(C_x)$, rotation $z_{rot}(C_x)$ and vibration $z_{vib}(C_x)$ for the cages and the clusters yield the combined partition function

$$Z(C_{60}) = z_{tr}(C_{60})z_{rot}(C_{60})z_{vib}(C_{60}),$$

$$Z(C_{60-x}) = z_{tr}(C_{60-x})z_{rot}(C_{60-x})z_{vib}(C_{60-x}) \text{ and}$$

$$Z(C_x) = z_{tr}(C_x)z_{rot}(C_x)z_{vib}(C_x).$$

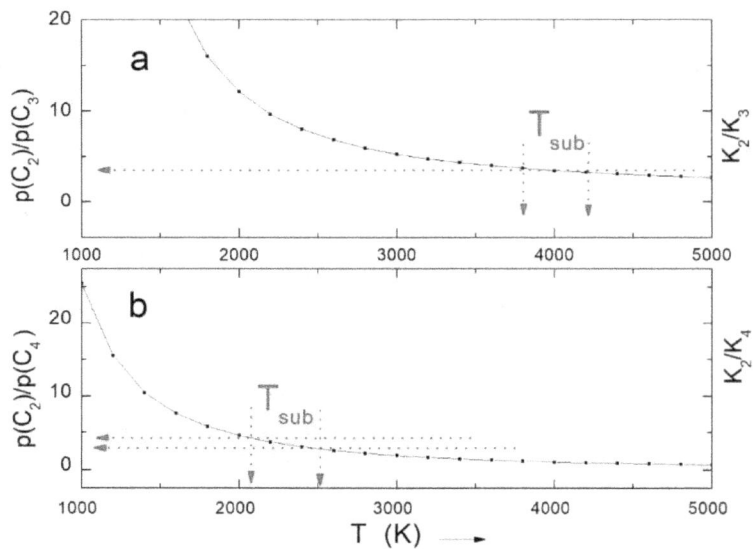

Fig. 7.4. Thermal and kinematical models for irradiated SWCNTs and C$_{60}$. (a) SWCNTs: Sublimation temperature $T_{Sub} \cong [(E_{xv} - E_{yv})/k)][\ln(p_y/p_x)]^{-1}$ is plotted as a function of the ratio of C$_2$ and C$_3$ from data in Fig. 7.3(a). The consistent ratio of p(C$_2$)/p(C$_3$)~3 yields T_{Sub}~4000K. (b) C$_{60}$-fullerite: T_{Sub} is calculated from the rate equation $K_x = Z(C_{60-x})Z(C_x)/Z(C_{60}) \, e^{-\frac{E(C_x)}{kT}}$ for the emissions of C$_2$ and C$_4$ from data in Fig. 7.3(b). Ratios of $K_2/K_4 \approx$ p(C$_2$)/p(C$_4$) are used to calculate T_{Sub}~2200 ± 200K.

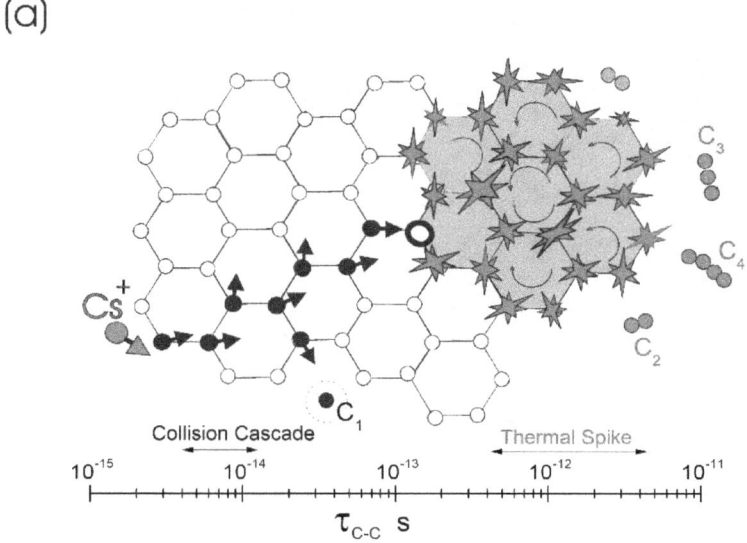

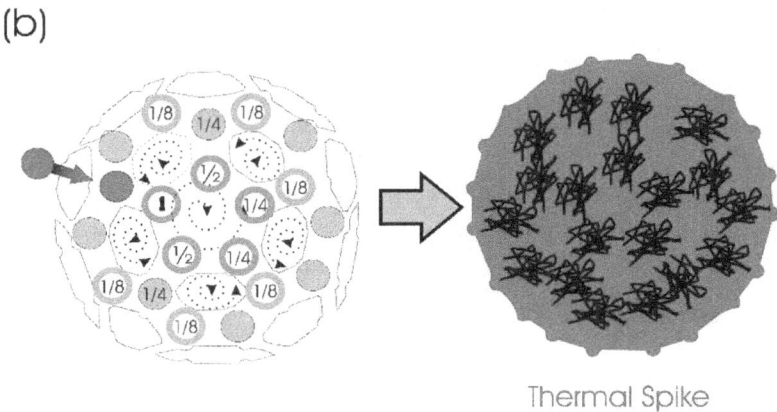

Fig. 7.5. Schematic representations of the thermal and SRS models. (a) In SWCNTs the initiation of collision cascades by energetic recoils is shown leading to thermal spikes that are characterized by the emission of monatomic and cluster emissions. (b) Irradiated C_{60} is shown to share the primary recoil energy with all atoms that leads to hot-explosive cage. In its initial stages, only C_2 and C_4 are emitted. The shrinking cage eventually explodes.

Out of the four emitted species C_1, C_2 C_3 and C_4, the ratio K_2/K_3 and K_2/K_4 are evaluated as a function of temperature T. The cross over with the experimentally determined ratios $p(C_2)/p(C_3)$ and $p(C_2)/p(C_4)$ determines the measured T_{Sub}.

In the case of fragmentation sequences demonstrated by irradiated-C_{60} in figure 8.3(c), the emerging profile of C_4 is different from those of the other emissions, showing stages of the ongoing processes in fragmenting cages. Average of the two values of T_{sub} are obtained from $p(C_2)/p(C_4)$ are $T_{sub} \sim 2200 \pm 200\ K$ for the irradiated C_{60}-fullerite.

Figure 7.5 presents the energy-dissipating, thermal modular descriptions of the linear-to-nonlinear transformation of $CC \rightarrow LTS$ in SWCNTs with associated time scales and fractal dimensions and the fragmenting, exploding C_{60} cages. These are compare with entropy-generating, representations of the evolution and involution of dissipative structures illustrated by the information-theoretic model.

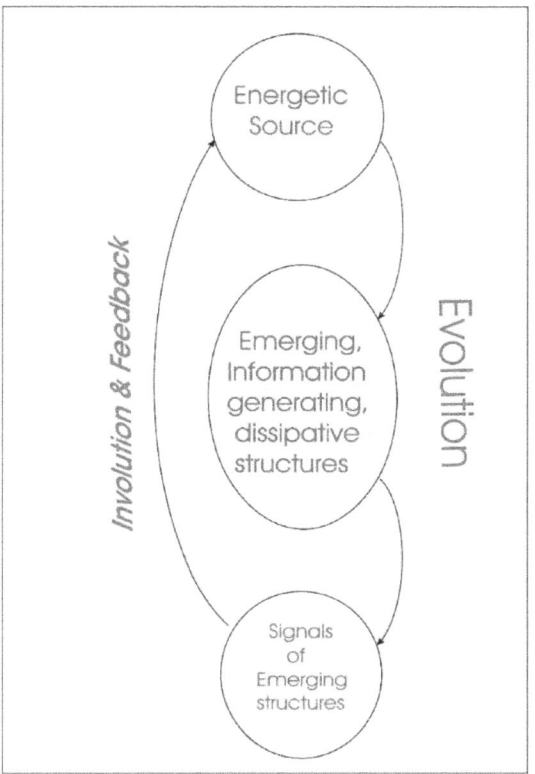

Fig. 7.6. The graphical summary of Chapter 7.

The evolutionary profile of dissipative structures (cascades, spikes, fragmenting and exploding cages) in irradiated SWCNTs and C_{60} cages is shown as the information-generating dynamic emergence that results from energetic input. The output signals in the form of sputtered carbon species, can be used as feedback for tailoring the emergence of the dynamical system (Cs^+ + C nanostructures) for the desired outputs, by adjusting the input parameters by involution and feedback.

The Self-Organizing Soot

Chapter 8

The Information manipulating Soot

8.1. The Cage→Cage transformations

Cage-fragmentation in a hot, grand canonical ensemble of fullerenes, can initiate the top-down shrinking sequences for the fullerenes $> C_{60}$, as discussed and illustrated in section 4.5 of Chapter 4. The C_x cage's transformation into a smaller cage with the emission of a C_2 molecule in the reaction $C_x \rightarrow C_{x-2} + C_2$ was suggested as the top-down, shrinking route. This results in the changes, of the configuration of the dynamical system composed of the fullerenes of various sizes and isomeric number densities, with each fragmentation step. The C_2-extrusion results in the decrease of the fullerene sizes. The reverse process can also occur by the C_2-ingestion. Together, these reactions were argued as the primary agents of the regenerative soot in Chapter 2. The fragmenting processes are the integral part of the information generating dynamical system with the mapping of an fullerene with x-atoms C_x onto an (x-2)-atom cage through the $cage \rightarrow cage + C_2$ transformations

$$f: C_x \rightarrow C_{x-2} + C_2 \qquad \text{eq. (8.1).}$$

The iterations of this cage-to-cage transformations, in equation (8.1) have been carried out for four sets of the self-organizing, fragmenting ensembles of the sets of fullerenes from C_{60} to C_{70} ($\sum_{60}^{70} C_x$), C_{60} to C_{80} ($\sum_{60}^{80} C_x$) and from C_{60} to C_{100} ($\sum_{60}^{100} C_x$). The summation notation for the ensembles of fullerenes C_x implies that the ensemble contains from C_{60} to the largest fullerene shown as the upper limit on the summation sign with the difference of a C_2. For example, $\sum_{60}^{70} C_x$ consist of the array of six fullerenes C_{60}, C_{62}, C_{64}, C_{66}, C_{68} and C_{70}.

Let us apply the SRS model developed in Chapter 6, to the case of the emergence of the Buckyball with icosahedral symmetry, out of the fragmenting, reforming ensembles of the fullerene cages. As in Chapter 4, we assume that the number densities of the fullerenes are in the ratios of their isomers. Therefore, majority of the isomers, of all of the fullerenes, have lower symmetries [3]. The larger cages have the exponentially increasing number of isomers described in section 4.2. Fullerenes' ability for mutual transformation, as discussed in Chapter 3-5, where the fullerenes cycle through the $cage \rightarrow cage$ transformations during their formative and fragmentation stages and for optimized cage-transformation conditions, the icosahedral C_{60} has been shown to emerge as the sole survivor. Fullerenes' self-organization in hot carbon vapor is treated here, as a dissipative dynamical system whose configuration changes with time. We evaluate the nonlinear interactions among the dynamical system's constituents and their mutual interdependence. Fragmenting, re-forming fullerenes and an evolving gas of C_2, as described in Chapter 4 and 5, are the constituents. The instantaneous, $p_x(\zeta)\ln(1/p_x(\zeta))$ and the total $\sum p_x(\zeta)\ln(1/p_x(\zeta))$ information generated by the constituents are calculated for the iterations of the mapping $f: cage \rightarrow cage + C_2$. From the information-theoretic profiles of the transformation of the entire fullerene ensembles into C_{60} and C_2, the fractal or Information dimensions of the fragmenting and evolving fullerenes are evaluated. The SRS model, developed in Chapter 6, will be employed to describe the conditions for the emergence of C_{60} as the end-directed emergence of the self-organizing, dissipative, dynamical systems of the fragmenting, transforming C cages.

The model and the data presented in this chapter is based on ref. [142].

8.2. The fragmenting fullerenes

Each fullerene C_x is mapped onto its fragmented components, a cage C_{x-2} and a C_2 molecule. The graphs in Fig. 8.1 are the iterative simulations of $f: C_x \rightarrow C_{x-2} + C_2$ from eq. (8.1). The iterations of cage→cage transformations are shown for the three such self-organizing, fragmenting dynamical systems-the ensembles of fullerenes designated as $\sum_{60}^{70} C_x$, $\sum_{60}^{80} C_x$ and $\sum_{60}^{100} C_x$.

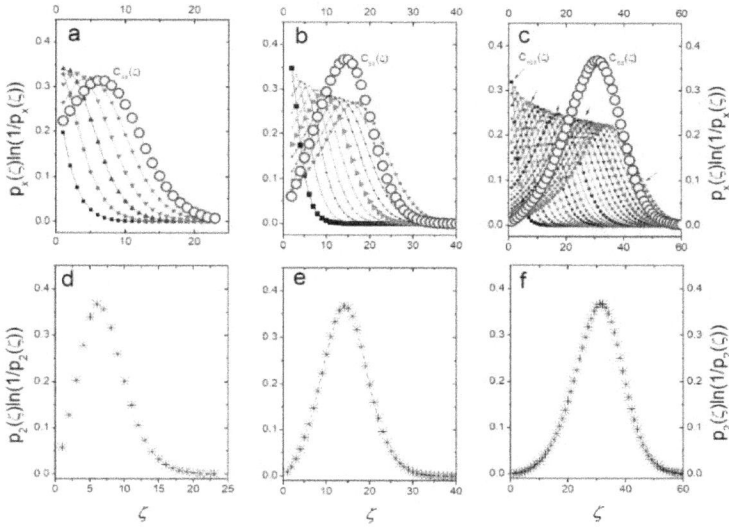

Fig. 8.1. **The fragmentation profile of the three ensembles of fullerenes $\sum_{60}^{70} C_x$, $\sum_{60}^{80} C_x$ and $\sum_{60}^{100} C_x$.** For each cage with x-carbon atoms, $p_x(\zeta)\ln(1/p_x(\zeta))$ is computed and plotted at every fragmentation step ζ. At fragmentation step $\zeta=0$, all cages start with their respective isomeric densities. Gradual build-up of C_{60} is visible from (a) to (c). The emerging information-theoretic profile of the C_2 gas $p_2(\zeta)\ln(1/p_2(\zeta))$ is shown for the three respective ensembles (d) to (f).

The range of fullerenes in each ensemble is from the lower to the higher limit of summation. In Figure 8.1(a), the results of the iterations for the ensemble of six fullerenes of the ensemble $\sum_{60}^{70} C_x$ are shown for each fragmenting step ζ. The normalized probability $p_x(\zeta)$ is evaluated for each fullerene C_x of the ensembles, taking the initial number densities from the isomer per fullerene data from ref [3]. Figure 8.1 has the instantaneous information $p_x(\zeta)\ln(1/p_x(\zeta))$ of all the cage-transformations calculated and plotted as a function of ζ. The total number of cages remains constant in $cage \rightarrow cage$ transformations.

8.3. The phase space of the self-organizing fullerenes

The sum of entropies of all cages at each successive fragmentation stages ζ was defined, in Chapter 6, as state of the dynamical system in eq. (6.2) $I_\zeta = \sum_x p_x(\zeta)\ln(1/p_x(\zeta))$. The orbit of I_ζ defines the phase space of the self-organizing ensembles of the fragmenting fullerenes.

The four sets of spectra of I_ζ versus ζ are the phase trajectories shown in Fig. 8.2(a). The four spectra represent the sum of entropies of all cages of the respective ensembles at each fragmentation stage ζ. Each of the I_ζ spectrum is the orbit of the whole of the ensemble during fragmentation. The phase trajectories of the dynamical system of the transformation of large fullerenes into the successively smaller ones represent the dynamic profile of the self-organization of fullerenes towards C_{60}. The dynamic pathways for the transition of the entire ensemble of fullerenes into the two gases of C_{60} and C_2 can be represented as

$$\sum_\zeta \sum_x C_x(\zeta) \rightarrow \sum_\zeta C_{60}(\zeta) + \sum_\zeta C_2(\zeta) \quad \text{eq. (8.2).}$$

Eq. (8.2) represents the entropic profile of the cage-to-cage transformations of the entire ensemble. The C_{60} can be seen emerging out in the entropic graphs. The emergence of C_{60} is the end-directed result of the cumulative fragmentations of all fullerenes into the next smaller one. All fullerene profiles, except that of the C_{60}, self-organize into the entropic curve of C_{60} as the net outcome of the dissipative, nonlinear dynamical system of ensembles of fragmenting, non-icosahedral fullerenes. We have used the same rates of fragmentation i.e., the @1/2 which implies half of all the fullerenes fragment and transform into the next smaller ones as indicated in the mapping scheme $C_x \rightarrow C_{x-2} + C_2 \rightarrow C_{x-4} + C_2 \rightarrow \cdots C_{60} + C_2$. The cumulative results were shown. These are the profiles of the fragmenting fullerenes.

The phase space profile of the shrinking and disappearing fullerenes and those of the emerging ones, as a function of the fragmenting steps are shown for the four ensembles from $\sum_{60}^{100} C_x$ to $\sum_{60}^{70} C_x$ in figure 8.2. Fig. 8.2(a) shows the sum of instantaneous information of all the cages from $x=60$ to 70, 80, 90 and 100, in the respective ensembles, represented as $I_\zeta = \sum_x p_x(\zeta) \ln(1/p_x(\zeta))$ at each fragmenting step ζ. Each data point represents the state of the entire ensemble. Fig. 8.2(b) plots the emerging, instantaneous $p_{60}(\zeta)\ln(1/p_{60}(\zeta))$ as a function of ζ entropic profiles for C_{60} for the 4 ensembles. We have included all cage transformations of the type $C_x \rightarrow C_{x-2} + C_2 \rightarrow C_{x-4} + C_2 \rightarrow \cdots C_{60} + C_2$. In Fig. 8.2(c) the instantaneous information trajectories of the evolving C_{60} gas $p_{60}(\zeta)\ln(1/p_{60}(\zeta))$ as a function of ζ.

The sum over all x and ζ of $p_x(\zeta)\ln(1/p_x(\zeta))$, is the entropic cost of the dynamical transition of the entire ensemble of fullerenes $\sum_x C_x(\zeta)$ into the two sets of the gases of C_{60} and C_2. The phase trajectories of the transformation of the ensembles, at each fragmentation step ζ represent the emerging profiles of the ensembles of self-organizing fullerenes. Due to the nature of the assumptions, the total number of cages remains constant in $cage \to cage + C_2$ transformations, only the number density of C_2 increases with each fragmentation step ζ.

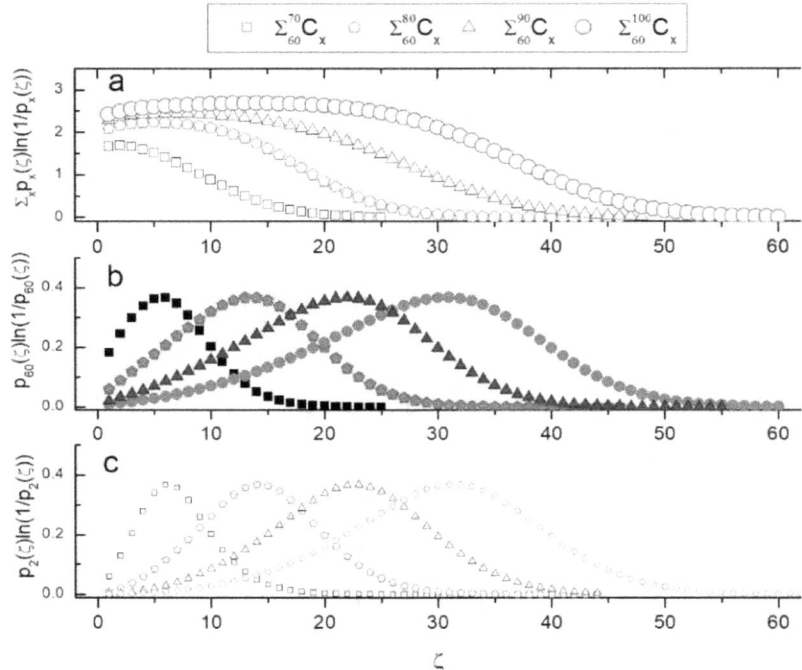

Fig.8.2. (a) **The phase space profiles of the shrinking and disappearing fullerenes.** The symbols for the four dynamical systems of the fullerene ensembles are shown in the inset. Each point of the respective phase trajectories represents the state of all the fragmenting and evolving cages. (a) Each data point represents the sum $\sum_x p_x(\zeta)ln(1/p_x(\zeta))$ of all the cages at each ζ. (b) The instantaneous information trajectories of the evolving C_{60} gas $p_{60}(\zeta)ln(1/p_{60}(\zeta))$ as a function of ζ. (c) Similarly, the evolution of the C_2 gas is shown $p_2(\zeta)ln(1/p_2(\zeta))$ at each fragmentation step ζ in (c).

The SRS model applied to the interconnected, information generating and sharing fullerene ensembles is further diagnosed by looking at the profiles of Kolmogorov function $ln(1/p_x(\zeta))$ and the instantaneous information $p_x(\zeta)ln(1/p_x(\zeta))$ for the two sets of fragmenting fullerenes belonging to the ensembles $\sum_{60}^{70} C_x$ and $\sum_{60}^{100} C_x$ in

Fig. 8.3.

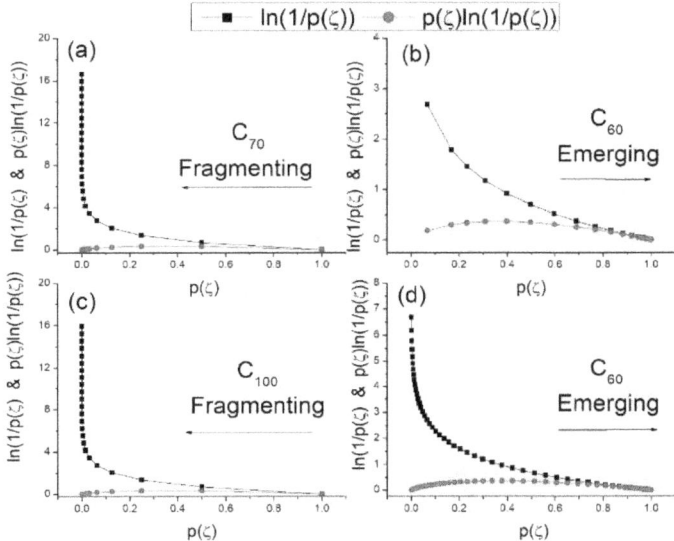

Fig. 8.3. The information theoretic parameters $ln(1/p_x(\zeta))$ and $p_x(\zeta)ln(1/p_x(\zeta))$ for the two ensembles are plotted for the two fullerenes; the largest and the emerging one. For $\sum_{60}^{70} C_x$ the fragmenting C70 is shown in (a) and for C60 in (b); one fragmenting and the other is the emerging fullerene. (c) is the fragmenting profile of C100 and (d) for C60 for the ensemble $\sum_{60}^{100} C_x$.

Figure 8.3 is the application of the SRS model developed for the interconnected, information generating and sharing two sets of fragmenting fullerenes belonging to the ensembles $\sum_{60}^{70} C_x$ and $\sum_{60}^{100} C_x$ operate as the information-sharing Boxes of Chapter 6. We have used the same rates of fragmentation i.e., the @1/2 which implies half of all the fullerenes fragment and transform into the next smaller ones as indicated in the mapping scheme $C_x \rightarrow C_{x-2} + C_2 \rightarrow C_{x-4} + C_2 \rightarrow \cdots C_{60} + C_2$. The cumulative results were shown in figures 8.1 and 8.2. Here the functions $\ln(1/p_x(\zeta))$ and $p_x(\zeta)\ln(1/p_x(\zeta))$ for only the two fullerenes C_{70} and C_{100} are plotted as a function of the probability $p_x(\zeta)$. These are the profiles of the fragmenting fullerenes. The emerging cage is C_{60} that is shown in the two sequences of transformations in 8.3(b) and (d).

The objective of Fig. 8.3 is to demonstrate that much higher information is generated by the fragmenting fullerenes C_{70} and C_{100}, in Fig. 8.3(a) and (c), for $p_x(\zeta) \leq 0.2$. The values of their instantaneous information $p_{70}(\zeta)\ln(1/p_{70}(\zeta))$ and $p_{100}(\zeta)\ln(1/p_{100}(\zeta))$ continue to increase for the decreasing number densities of C_{70} and C_{100}. The opposite is true for the emerging C_{60}. This aspect of the increasing information as a function of the probability is explained by the consequent increasing function $\ln(1/p_x(\zeta))$.

8.4. Fractal dimension of the self-organizing cages

From the graphs of $p_x(\zeta)\ln(1/p_x(\zeta))$ against ζ plotted in Fig. 8.1, the information or the Shannon entropy for the shrinking ($>C_{60}$) and the accumulating (C_{60}) fullerenes is calculated from I_x in eq. (6.2). It is the area under the entropic profiles or the trajectories for each of the fullerenes and the associated C_2s. Information I_x or the Shannon entropy of the self-organizing fullerenes for the four ensembles $\sum_{60}^{70} C_x$, $\sum_{60}^{80} C_x$, $\sum_{60}^{90} C_x$ and $\sum_{60}^{100} C_x$ is shown in Fig. 8.4(a). Applying the SRS model to the fullerenes belonging to any of the four ensembles, starting with their initial, pre-fragmentation densities yield their respective information. Their densities are reduced by half at every fragmentation step. The largest fullerenes fragment away within 10-15 ζ stages. The smallest one, C_{60}, being the recipient of all of the cage- transformations, has the entire fragmenting ensemble as its basin. Therefore, it has the largest number of accumulation stages ~ the maximum number of ζ.

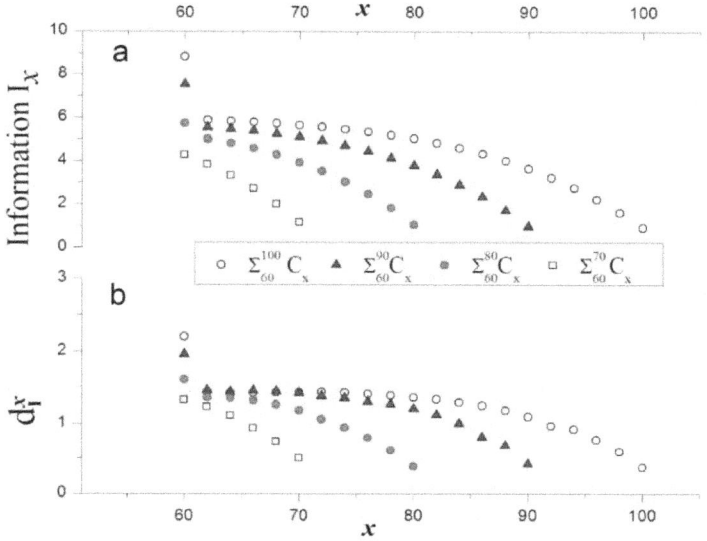

Fig. 8.4. Fractal description of self-organizing fullerenes. (a). The Shannon entropy or the Information I_x from equation (2) is plotted for all transforming cages of the four dynamical system shown in the inset. Information maximizes for all the ensembles as the fragmenting cages shrink towards C_{60}. (b). Fractal dimensions d_I^x are evaluated using I_x and plotted for every constituent fullerene by using equation (3). **Inset:** the symbols for the fullerenes belonging to the four ensembles are shown.

Total Information I_x is obtained by summing the instantaneous information $p_x(\zeta)\ln(1/p_x(\zeta))$ at each fragmentation step for every cage, over all ζ. The calculated total information follows the trend $I_x < I_{x-2} < I_{x-4} ... < I_{60}$ for the fullerenes of the four ensembles. This pattern is clearly visible in Fig. 7.4(a). The cumulative results for the ensemble show the fragmentation pattern of fullerenes. The sequences of all cage fragmentations lead to the enrichment of a single species i.e. C_{60}. The largest ensemble also has the highest I_x for C_{60}.

In Fig. 8.4(b), the fractal dimensions d_I^x of the constituent fullerenes are calculated by using eq. (6.3). The fractal dimension varies for all fullerenes of the ensemble. It follows the trend shown by the corresponding information I_x in 8.4(a). The dynamical system with wider range of x like $\sum_{60}^{100} C_x$ yield higher fractal dimension for C_{60}. Fractal dimension d_I^{60} for the accumulating C_{60} gas increases with the size of the ensemble. This trend is a natural consequence of the availability of the increasing numbers of larger (> C_{60}) cages. Therefore, due to higher abundance for the larger fullerenes with lower cage formation energies, the starting ensemble with successively larger fullerenes will always have increasing number densities for cages > C_{60} and hence the largest fractal dimension d_I^{60} for C_{60}.

Similarly, the fractal dimension of the accumulating C_2-gas can be obtained from $d_I^2 = \sum_\zeta p_2(\zeta) \ln(1/p_2(\zeta)_2 / \ln(1/\zeta)$. Fractal dimension d_I^x for each fullerene shown in Fig. 8.3(b) is obtained from the respective information I_x. C_{60} has the highest fractal dimension $d_I^{60} > d_I^{x>60}$. It demonstrated the end-directed emergence of the dynamical system of the self-organizing ensembles of fullerenes.

Information is shown to be the crucial link between the phase space of a self-organizing dynamical system and the fractal dimensions of its constituents. The data obtained for the four dynamical systems are tabulated as Table 81. It has the cumulative Information for all of the cages $\sum_\zeta \sum_x p_x(\zeta) \ln(1/p_x(\zeta)$, the evolving gas of C_{60}s $\sum_\zeta p_{60}(\zeta) \ln(1/p_{60}(\zeta))$ and for the C_2s gas as $\sum_\zeta p_2(\zeta) \ln(1/p_2(\zeta)_2$. The fractal dimensions d_I^{60} and d_I^2 of the accumulating gases of C_{60} and C_2 have been calculated and displayed for the four ensembles. The ratio of the total information of the fragmenting fullerenes and the C_2 gas to that of the emerging C_{60}s is defined as $i_{60}^\Sigma \equiv I_{total}/I_{60}$. It has numerical values 6.86, 8.86, 10.53 and

12.77 for the four ensembles $\sum_{60}^{70} C_x$, $\sum_{60}^{80} C_x$, $\sum_{60}^{90} C_x$ and $\sum_{60}^{100} C_x$. Similarly the fractal dimension d_I^{60} is evaluated for the fragmenting fullerenes of the dynamical systems $\sum_\zeta \sum_x C_x(\zeta) \to \sum_\zeta C_{60}(\zeta) + \sum_\zeta C_2(\zeta)$ where the larger cages transform into C_{60}. It yields successively increasing the information dimension $d_I^{60} = 1.13$ to 2.1 for the 4 dynamical systems.

Table 8.1. Information for all of the cages of the four ensembles and the emerging gases of C_{60} and C_2 are tabulated. These are shown as $\sum I_x \equiv \sum_\zeta \sum_x p_x(\zeta) \ln(1/p_x(\zeta))$, $I_{60} = \sum_\zeta p_{60}(\zeta) \ln(1/p_{60}(\zeta))$ and $I_2 = \sum_\zeta p_2(\zeta) \ln(1/p_2(\zeta))$. The fractal dimension d_I^{60} of the emergent structure C_{60} and the C_2 gas are tabulated.

	$\sum_{60}^{70} C_x$	$\sum_{60}^{80} C_x$	$\sum_{60}^{90} C_x$	$\sum_{60}^{100} C_x$
$\sum I_x$	23.34	50.79	79.42	112.8
I_{60}	3.4	5.73	7.54	8.83
d_I^{60}	1.13	1.61	1.9	2.1
I_2	2.66	4.88	6.39	7.67
d_I^2	0.99	1.37	1.68	1.87

8.5. The SRS model of the self-organizing ensembles of fullerenes

1. A model for the fullerenes in hot carbon vapour environment where the cage formation and fragmentation takes place, was developed in which the self-organizing fullerene ensembles are modeled as dissipative dynamical systems.

2.	Starting with the number densities in the respective ratios of the isomeric possibilities of the fullerenes in the ensemble, each cage is subjected to the internal cage-fragmentation forces described in Chapters 3 and 5.

3.	The larger cages transform into the successive smaller ones and are mapped by $f: C_x \to C_{x-2} + C_2$ from eq. (8.1). From the density variation of the cages, the probabilities of the entire $cage \to cage + C_2$ transformations are calculated at each fragmentation step ζ.

4.	The Information or the Shannon entropy summed over the instantaneous entropic profiles of all fragmenting stages is first tabulated and then used to calculate the fractal dimension of all constituents of the dynamical systems.

5.	We have shown that the dissipative dynamical system of fragmenting and transforming cages has an end-directed emergence towards C_{60}. It acts as the sink whose basin is the entire ensemble.

6.	Information is the crucial link between the phase space of a dissipative dynamical system and the fractal dimensions of its constituents. It helps us to identify the processes of self-organization.

Figure 8.5 is the graphical summary of the SRS model applied to the self-organizing fullerenes. An ensemble $\sum_{60}^{80} C_x$ consisting of the 11 fullerenes C_{60}, C_{62}, C_{64}, C_{66}, C_{68}, C_{70}, C_{72}, C_{74}, C_{76}, C_{78} and C_{80} transform into C_{60} via $\sum_{60}^{80} C_x \to \sum C_{60} + \sum C_2$. Feedback mechanisms are highlighted. Involution and feedback of material and information are the essential elements of self-organization.

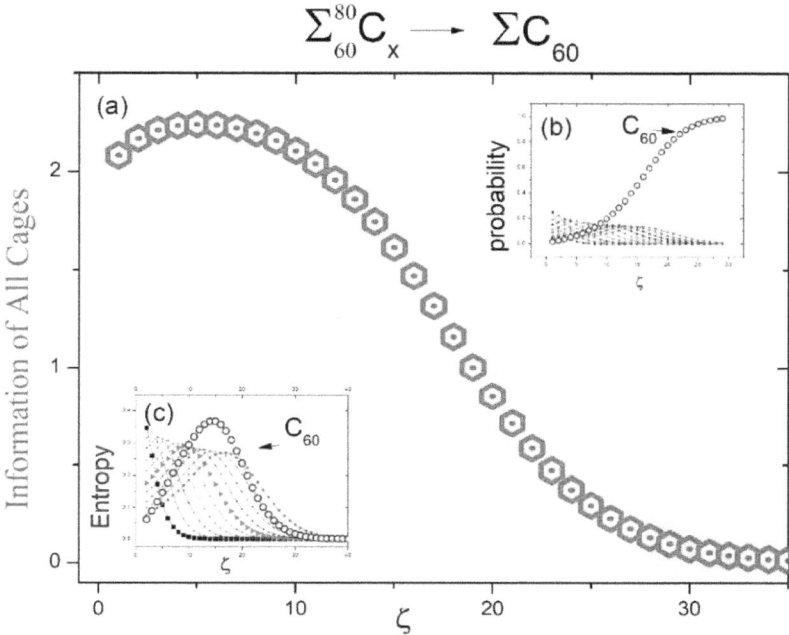

Fig. 8.5. The graphical summary of Chapter 8.

The emergence of C_{60} out of the ensemble of fragmenting and reforming fullerenes is shown as the end-directed emergence of the dynamical system $\sum_{60}^{80} C_x \to \sum C_{60}$. It describes graphically the generation of information, instantaneous and total, by the ensemble of fragmenting fullerenes. The initial number densities of ensemble's fullerenes are assumed to be proportional to their isomeric densities. The non-icosahedral fullerenes are subjected to the internal cage-fragmentation forces described in Chapters 3 and 5. The ensemble is embedded in the hot sooty environment of the condensing carbon vapour. The model assumes fragmentation of half of all fullerenes at each fragmentation stage ζ. The larger fragmenting cages populate the smaller cage densities.

The Self-Organizing Soot

LIST OF REFERENCES

[1]   R.E.Smalley, Nobel Lecture, 1996.

[2]   H.W. Kroto, J.R. Heath, C.S. O'Brien, R.F. Curl & R.E. Smalley, Nature (London) **318**, 162 (1985).

[3]   P.W. Fowler, D.E. Manolopoulos, *An Atlas of Fullerenes* (Oxford: Clarendon, 1995).

[4]   S. Iijima, Nature (London) **354**, 56 (1991).

[5]   W. Krätschmer, L.D. Lamb, K. Fostiropoulous & D.R. Hoffman, Nature (London) **347**, 354 (1990).

[6]   M.S. Dresselhaus, G. Dresselhaus, P.C. Elkund, *Science of Fullerenes and Carbon Nanotubes: Their Properties and Applications* (Academic, San Diego, 1996).

[7]   M.N. Akhtar, B. Ahmad, S. Ahmad, Phys. Lett. A **234**, 367 (1997).

[8]   S. Ahmad, M.N. Akhtar, A. Qayyum, B. Ahmad, K. Babar, W. Arshad, Nucl. Instrum. Methods Phys. Res. B **122**, 19 (1997).

[9]   A. Qayyum, B. Ahmad, M.N. Akhtar, S. Ahmad, Eur. Phys. J. D **3**, 267 (1998).

[10] S. Javeed, S. Ahmad (2021) *Experimental and Theoretical Aspects of the Fragmentation of Carbon's Single- and Multiwalled Nanotubes*. In: Abraham J., Thomas S., Kalarikkal N. (eds) Handbook of Carbon Nanotubes. Springer, Cham. https://doi.org/10.1007/978-3-319-70614-6_71-1.

[11] R.A. Middleton, Nucl. Instrum. Methods, **144**, 373 (1977).

[12] R.E. Honig, J. Chem. Phys. **22**, 126 (1954); ibid 1610 (1954).

[13] J. Berkowitz, W.A. Chupka, J. Chem. Phys. **40**, 2735 (1964).

[14] A.G. Whittaker, P.L. Kintner, Carbon, **7**, 414 (1969).

[15] S. Yang, K. J. Taylor, M. J. Craycraft, J. Conceicao, C. L. Pettiette, O. Chesnovsky, and R. E. Smalley, Chem. Phys. **144**, 431 (1988).

[16] F. Lépine, A. R. Allouche, B. Baguenard, Ch. Bordas, and M. Aubert-Frécon, J. Phys. Chem. A **106**, 31, 7177–7183 (2002).

[17] P. Sigmund in *Sputtering by particle bombardment,* vol. I (ed). R. Behrisch (Springer, Berlin, 1981).
[18] M. W. Thompson, R. Nelson, Philos. Mag. **7**, 2015 (1962).
[19] G. Vineyard, Radiat. Eff. **29**, 245 (1976).
[20] R. Kelly, Radiat. Eff. **32**, 91(1977).
[21] P. Sigmund, C. Claussen, J. Appl. Phys. **52**, 990 (1981).
[22] S. Ahmad, B. Farmery, M. W. Thompson, Nucl. Instrum. Meth. **170**, 327(1980); Philos. Mag. A **44**, 1387(1981).
[23] H. M. Urbassek, K. T. Waldeer, Phys. Rev. Lett. **67**, 105 (1991).
[24] M. Kerford, R. P. Webb, Carbon, **37**, 859 (1999).
[25] W. O. Hofer in *Sputtering by particle bombardment,* vol. III (eds.) R. Behrisch, K.Wittmaack, (Springer, Berlin 2007).
[26] S. Javeed, S. Ahmad, Philos. Mag. **97**, 1436 (2017).

[27] A. V. Krasheninnikov, P. O. Lehtinen, A.S. Foster, P. Pyykkö, R.M. Nieminen, Chem. Phys. Lett. **418**, 132(2006).
[28] A. J. Lu and B. C. Pan, Phys. Rev. Lett. 92, 105504 (2004).
[29] S. Berber, A. Oshiyama, Phys. Rev. B **77**, 165405 (2008).
[30] A. V. Krasheninnikov, and F. Banhart Nature Materials, **6** 723 (2007).

[31] A.V. Kresheninnikov, K. Nordlund, J. Appl. Phys. **107**, 071301 (2010).

[32] N. Bohr, Dan. Vid. Selsk Mat. Fys. Medd. **18**, No. 8 (1948).

[33] W. Eckstein, Nucl. Instrum. Methods Phys. Res. B **27**, 78 (1987).

[34] M.W. Thompson, Defects and Radiation Damage in Metals, (Cambridge University Press, Cambridge, 1969).

[35] S. Ahmad, T. Riffat, Nucl. Instrum. Meth. Phys. Res. B **152**, 506 (1999);

S. Ahmad, A. Qayyum, M.N. Akhtar, T. Riffat, Nucl. Instrum. Meth. Phys. Res. B **171**, 551 (2000).

[36] A. Qayyum, M.N. Akhtar, T. Riffat, S. Ahmad, Appl. Phys. Lett. **75**, 4100 (1999);

S.Ahmad, B. Ahmad, A. Qayyum, M.N. Akhtar, Nucl. Instrum. Meth. Phys. Res. A **452**, 370 (2000).

[37] S. Ahmad, Review: *The Spectroscopy of the regenerative soot*, Eur. Phys. J. D **18**, 309 (2002).

[38] D. Ugarte, Nature **359**, 707 (1992).

[39] F. Banhart, Rep. Prog. Phys. **62**, 1181 (1999).

[40] D. Fink, K. Ibel, P. Goppelt, J.P. Biersack, L. Wang, M. Baher, Nucl. Instrum. Meth. Phys. Res. B **46**, 342 (1990).

[41] G. Brinkmalm et al, Chem. Phys. Lett. **191**, 345 (1992).

[42] L. Chadderton et al Nucl. Instrum. Meth. Phys. Res. B **91**, 71 (1994).

[43] S. Ahmad, M.N. Akhtar, Appl. Phys. Lett. 78, 1499 (2001).

[44] S. Ahmad, Eur. Phys. J. AP **5**, 111 (1999); Phys. Lett. A **261**, 327 (1999).

[45] NIST Atomic Spectra Database (ADS) Data at http://physics.nist.gov/

[46] J.F. Ziegler, J.P. Biersack, U. Littmark, *The Stopping and Range of Ions in Solids* (Pergamon Press, New York, 1985), Vol. 1.

[47] N. Falk, Improved *Hollow Cathode lamps for Atomic Spectroscopy*, edited by C. Sergio (Chichester, Ellis Horwood, 1985).

[48] W. Lotz, ApJ **14**, 207 (1967).

[49] E. Bohm-Vitense, *Introduction to Stellar Atmospheres* (Cambridge University Press, Cambridge, 1993), Vol 2.

[50] S. Ahmad, B. Ahmad, T. Riffat, Phys. Rev. E **64**, 026408

(2001).

[51] K. Raghavachari, J.S. Binsky, in *Physics and Chemistry of Small Clusters*, NATO ASI series B, (eds) P. Jena et al. (Plenum Press, New York, 1987) Vol. **158**, p. 317.

[52] S.W. McElvany, B.I. Dunlop, A. O'Keefe, J. Chem. Phys. **86**, 715 (1987).

[53] P.P. Radi, T.L. Bunn, P.R. Kemper, M.E. Molchn, M.T. Bowers, J. Chem. Phys.**88**, 2809 (1988).

[54] S.A. Janjua, *Study of physical mechanisms of regenerative sooting discharges*, PhD Thesis, PIEAS, Islamabad, (2007).

[55] S. Ahmad, Nanotechnology **16**, 1739 (2005).

[56] H.W. Kroto, Science **242**, 1139 (1988).

[57] A. Thess A et al Science **273**, 483 (1996).

[58] C. Journet et al Nature (London) **388**, 756 (1997).

[59] B. Bhushan, (ed) *Springer Handbook of Nanotechnology* (Springer, Berlin, 2004).

[60] G.G. Tibbetts, J. Cryst. Growth **66**, 632 (1984).

[61] T.G. Schmalz, W.A. Seitz, D.J. Klein, G.E. Hite, J. Am. Chem. Soc. **110** 1113 (1988).

[62] T. Tersoff, Phys. Rev. B **37**, 6991 (1988); Phys. Rev. B **46**, 15546 (1992).

[63] G.B. Adams, O.F. Sankey, J.B. Page, M. O'Keeffe, D.A. Drabold, Science **256**, 1792 (1992).

[64] K.H. Bennemann, D. Richardt, J.L. Moran-Lopez, R. Kerner, K. Penson, Z. Phys. D **29**, 231 (1994).

[65] D. Bakowies, W. Thiel, J. Am. Chem. Soc. **113**, 3704 (1991).

[66] A. Maiti, C.J. Brabec, J. Bernholc, Phys. Rev. Lett. **70**, 3023 (1993).

[67] T. Whitten, H. Li, Eur. Phys. Lett. **23**, 5 (1993).

[68] X. Xiaoyu, L. Ji-xing, O. Zhong-can, Mod. Phys. Lett. B **9**, 1649 (1995).

[69] B.T. Kelly, *Physics of Graphite* (Applied Sciences Publishers, London 1981).

[70] L.D. Landau, E.M. Lifshitz, *Theory of Elasticity* 3rd Ed. (Pergamon, London, 1986).

[71] S. Timoshenko, S.W. Krieger, *Theory of Plates and Shells* 2nd Ed. (McGraw Hill, New York, 1959).

[72] D. Ugarte, Nature (London) **359,** 707 (1992); Eur. Phys. Lett. **22,** 45 (1993).

[73] C.S. Yannoni, R.D. Johnson, G. Meijer, D.S. Bethune, J.R. Salem, J. Phys. Chem. **95**, 9 (1991).

[74] J.P. Lu, X.P. Li, R.P. Martin, Phys. Rev. Lett. **68**, 1551 (1992).

[75] G. Overney, W. Zhong, D. Tomanek, Z. Phys. **27**, 93 (1993).

[76] E. Hernandes, C. Goze, P. Bernier, A. Rubio, Phys. Rev.Lett. **80**, 4502 (1998).

[77] X. Blasé, A. Rubio, S.G. Louie, M.L. Cohen, Eur. Phys. Lett. **28**, 335 (1994).

[78] D.H. Robertson, D.W. Brenner, J.W. Mintmire, Phys.Rev. B **45**, 12592 (1992).

[79] J.P. Lu, Phys. Rev. Lett. **79**, 1297 (1997).

[80] H.W. Kroto, K. McKay, Nature **331**, 328 (1998).

[81] S. Ahmad, K. Yaqub, A. Ashraf, Eur. Phys. J. D. **67**, 51 (2013).

[82] R.F. Curl, R.E. Smalley, Sci. Am. **265**, 32 (1991).

[83] J.R. Heath, ACS Symp. Ser. **481**, 1 (1991).

[84] M. Endo, H. W. Kroto, J. Phys. Chem. **96**, 6941 (1992).

[85] B. I. Dunlap, Int. J. Quantum Chem. **64**, 193 (1997).

[86] D.H. Robertson, J. Phys. Chem. **96**, 6133 (1992).

[87] T. Wakabayashi, Y. Achiba, Chem. Phys. Lett. **190**, 465 (1992).

[88] S. Mauyama, Y. Yamaguchi, Chem. Phys. Lett. **286**, 343 (1998).

[89] S. Irle, G. Zheng, Z. Wang, K. Morokuma, J. Phys. Chem. B **110**, 14531 (2006).

[90] S.D. Khan, S. Ahmad, Nanotechnology **17**, 4654 (2006).

[91] L.D. Landau, E.M. Lifshitz, *Statistical Mechanics* 3rd Ed. (Butterworth, Oxford, 1980).

[92] A.J. Stone, D.J. Wales, 1986 Chem. Phys. Lett. **128**, 501 (1986).

[93] J.P. Lu, W. Wang, Phys. Rev. B **49** 11421 (1994).

[94] E. Pasqualini, Phys. Rev. B 56 7751 (1997).

[95] D. Tomanek, M.A Schlutur, Phys. Rev. Lett. 67 2331 (1991).

[96] K.H. Bennemann, D. Richardt, J.L. Moran-Lopez, R. Kerner, K. Penson, Z. Phys. D **29**, 231 (1994).

[97] R.C. Haddon, L.E. Brus, K. Rangavachari, Chem. Phys. Lett. **131**, 165 (1986).

[98] D.P. Clougherty, X. Zhu, Phys. Rev. A **56**, 632 (1997).

[99] R.C. Haddon, Science **261**, 1545 (1993).

[100] M. A. Thompson, J. Phys. Chem. **100**, 14492 (1996).

[101] M.J. Drewer, W. Thiel, J. Am. Chem. Soc. **99**, 4899 (1977).

[102] S, Ahmad, S.D. Khan, S. Manzoor, Nanotechnology **17**, 1686 (2006).

[103] L. Pauling, J. Chem. Phys. **4**, 673 (1936).

[104] O. Schmidt, Z. phys. Chem. B **39**, 59 (1938); **42**, 83 (1939).

[105] J.R. Platt, J. Chem. Phys. **17**, 484 (1949).

[106] M.W. Nathans, J. Chem. Phys. **20**, 741 (1952).

[107] J.S. Griffith, J. Chem. Phys. **21**, 174 (1953).

[108] K. Fukui, T. Yonezawa, C. Nagata, and H. Shingu, J. Chem. Phys. **22**, 1433 (1954).

[109] C.A. Coulson, Proc. Phys. Soc. (London) A **66**, 652 (1953).

[110] W.T. Simpson, J. Chem. Phys. **17**, 1218 (1949).

[111] H.H. Jaffe, J. Chem. Phys. **20**, 1646 (1952).

[112] K. Rudenberg, C.W. Scherr, J. Chem. Phys. **21**, 1565 (1953).

[113] A.A. Frost, J. Chem. Phys. **23**, 310 (1955).

[114] N.S. Ham, K. Rudenberg, J. Chem. Phys. **29**, 1199 (1958).

[115] D. Ugarte, Nature (London) **359**, 707 (1992); Europhys. Lett. **22**, 45 (1993).

[116] M.S. Zwanger, F. Banhart, Philos. Mag. B **72**, 149 (1995).

[117] T. Cabioch, J.P. Riviere, J. Delfond, J. Mater. Sci. **30**, 4787 (1995),

[118] S. Ahmad, J. Chem. Phys. **116**, 3396 (2002).

[119] S. H. Strogatz, *Nonlinear dynamics and chaos* (Perseus Books, Reading, 1994).

[120] K. T. Alligood, T. D. Sauer, J. A. Yorke, Chaos-*An Introduction to Dynamical Systems*, (Springer-Verlag, New York, 1997).

[121] P. Berge, Y. Pomeau, C. Vidal, *Order within Chaos* (John Wiley & Sons, New York, 1984).

[122] G. Nicolis, I. Prigogine, *Self-organization in non-equilibrium systems* (John-Wiley & Sons, New York, 1977).

[123] I. Prigogine, I. Stengers, *Order Out of Chaos, Man's New Dialogue with Nature* (Bantam Books, New York, 1984).

[124] D. Kondepundi, B. Kay, J. Dixon, Chaos **27**, 104607 (2017).

[126] H. Haken, Rev. Mod. Phys. **47**, 67 (1975).

[126]   J. Skar, Phil Trans R Soc Lond A **361**, 1049 (2003).

[127]   J. D. Halley, D. A. Winkler, Complexity **13**, 10 (2008).

[128]   C. E. Shannon, *A Mathematical Theory of Communication*, Bell Systems Tech. J. **27**, 379 (1948).

[129]   S. Ahmad, Chaos **28**, 123125 (2018).

[130]   B. B. Mandlebrot, *The Fractal Geometry of Nature* (W. H. Freeman & Co., New York, 1977).

[131]   K. J. Falconer, *Fractal geometry* (John Wiley & Sons, Chichester 1990).

[132]   G. P. Cherepanov, A. S. Balankin, V. S. Ivanova, *Fractal fracture mechanics—A review*, Eng. Frac. Mech. **51**, 997 (1995).

[133]   S. Ahmad, M. S. Abbas, M. Yousuf, S. Javeed, S. Zeeshan, K. Yaqub, Eur. Phys. J. D, **72**, 70 (2018).

[134]   A. Renyi, Acta Math. Acad. Sci. Hungaria **10**, 193 (1959).

[135]   S. Kullback, R. A. Leibler, Ann. Math. Stat. **22**, 79 (1951).

[136]   A. N. Kolmogorov, *Information theory and the theory of algorithms. Selected works, Vol. 3* (Kluwer, Dordrecht, 1993).

[137]   T. M. Cover, J. A. Thomas, *Elements of Information Theory* (John Wiley & Sons, New York, 1991).

[138]   S. Javeed, *Synthesis and Fragmentation of Carbon Clusters and Nanotubes,* PhD Thesis, PIEAS, Islamabad (2014).

[139]   S. Zeeshan, *The Patterns and Routes of Fragmentation of the Irradiated C_{60} Fullerite and Powder*, PhD Thesis, PIEAS, Islamabad (2015).

[140]   A. Ashraf, S. Javeed, S. Zeeshan, K. Yaqub, S. Ahmad, Chem. Phys. Lett. **707**, 144 (2018).

[141]   D. A. McQuarrie, *Statistical Mechanics* (University Science Books, Suasalato, 2003).

[142]   S. Ahmad, Chem. Phys. Lett. **713**, 52 (2018).

The Self-Organizing Soot

ABOUT THE AUTHOR

Shoaib Ahmad has doctorate in physics. He worked and taught physics at British and Pakistani universities and research institutions. Besides publishing in physics journals, he writes on energy, social and cultural issues for the non-specialists, in Urdu and English. In recent years he has extended his experimental research into self-organizing, information generating nanostructures to developing mathematical models that can be applied to the emerging social and cultural dynamical systems. In this book, his information-theoretic Source-Reservoir-Sink (SRS) model, based on the probability distributions of the emerging, dissipative structures and events, is shown to generate comprehensive profiles of the energy dissipating, information-generating events and epidemics.